Studies in Computational Intelligence 425

Editor-in-Chief

Prof. Janusz Kacprzyk
Systems Research Institute
Polish Academy of Sciences
ul. Newelska 6
01-447 Warsaw
Poland
E-mail: kacprzyk@ibspan.waw.pl

For further volumes:
http://www.springer.com/series/7092

Zbigniew Les and Magdalena Les

Shape Understanding System – Knowledge Implementation and Learning

Springer

Authors
Prof. Dr. Zbigniew Les
The Queen Jadwiga Research Institute
of Understanding
Toorak, VIC
Australia

Magdalena Les
The Queen Jadwiga Research Institute
of Understanding
Toorak, VIC
Australia

ISSN 1860-949X e-ISSN 1860-9503
ISBN 978-3-642-29696-3 e-ISBN 978-3-642-29697-0
DOI 10.1007/978-3-642-29697-0
Springer Heidelberg New York Dordrecht London

Library of Congress Control Number: 2012937301

Printed on acid-free paper

Springer is part of Springer Science+Business Media (www.springer.com)

Dedication

This book is dedicated to our Patron St. Jadwiga Queen of Poland

Introduction

This book presents the selected results of research on the further development of the shape understanding system, the aim of which is to build the understanding/thinking machine. The shape understanding system that is an implementation of the shape understanding method was described in our previous book titled "Shape Understanding System: the first steps toward the visual thinking machines" [1]. This is the second book that presents the results of research in the area of thinking and understanding, carried out by authors in the newly founded the Queen Jadwiga Research Institute of Understanding. In the first book [1], a brief description of philosophical investigations of topics connected with understanding and thinking was presented. The shape that is the main perceptual category of thinking process and the important visual feature of the perceived world was briefly described in the first Section of the first Chapter. In the next Section, the different problems connected with understanding investigated by philosophers such as Locke or Berkeley were described. The relation between understanding and thinking was discussed in the following Section of the first Chapter. The last Section included the short description of the shape understanding system. In Chapter 2 the concept of shape classes that are regarded as the basic perceptual categories were presented. Shape classes are represented by their symbolic names. Each class is related to each other and, based on these classes, there is relatively easy to establish the 'perceptual similarity' among perceived objects. In Chapter 3 the description of the reasoning process that leads to assigning the perceived object to one of the shape classes was described. Assignment of the object to one of the general classes is based on the specific reasoning process. As the result of the reasoning process, an examined object is assigned to one of the shape classes where each class is represented by its symbolic name. The symbolic name is used to find the visual concept and next, to assign a perceived object into one of the ontological categories. In Chapter 4 the new hierarchical categorical structures of the different categories of the visual objects were presented. The categorical chains that represent the visual categories are applied to interpret the perceived object as the member of one of the ontological categories: the figure category, the sign category, the letter category or the real world object category. In Chapter 5 examples of the visual reasoning processes that can be considered as the special kind of the thinking processes were presented. The thinking process is regarded as the continuous computational activity that is triggered by perception of a new object, an 'inner object' or a task given by the user. Thinking can lead to solving a problem where there is only one solution (e.g. the visual intelligence test) or solving a problem where there are many possible solutions (e.g. designing the tools). In the first book [1], the focus was on thinking that leads to solving a problem that has only one solution.

In Chapter 1 of this book some aspects of human learning that are related to the newly introduced concept of the knowledge implementation are described. In Chapter 2 a short survey of literature, on the vast topic concerning learning by machine, is presented. In Chapter 3, the knowledge implementation is defined in the context of both human learning and machine learning. In Chapter 4 the selected issues connected with learning and understanding are described in the context of the newly introduced concept of the knowledge implementation. The relations between understanding and learning are also discussed. In Chapter 5 the shape understanding method is presented. The shape that is the main perceptual category of thinking process and the important visual feature of the perceived world is also briefly described in this Chapter. In Chapter 6 categories of the visual objects are described. This description is based on the material presented in our first book [1]. In Chapter 7 the theoretical framework of the knowledge implementation method is presented. In Chapter 8 the knowledge implementation approach that is used for learning knowledge and skills of the different categories of objects is presented. The theoretical framework of the knowledge implementation presented in Chapter 7 refers to two main modes in which SUS operates, namely, the learning and understanding mode. Learning and understanding are complementary processes. The SUS ability to understand depends on the effectiveness of learning process and learning of the new knowledge depends on the SUS ability to understand. The knowledge implementation is based on the assumption that a system, in order to be able to understand, needs to learn and learned knowledge needs to be fully understood. The knowledge implementation is concerned with learning of the visual and non-visual knowledge from the different categories of objects. The category of the visual objects that is connected with learning of the visual knowledge is described in Chapter 6. The category of the sensory objects and the category of text objects are related to the category of the visual objects and these categories are presented in Chapter 7. These categories are described in the context of learning and understanding of the visual object, the sensory object and the text object. Proposed new learning methods that are part of the knowledge implementation approach are designed to learn knowledge of an object from the selected categories such as the category of visual objects, the category of sensory objects and the category of text objects. In Section 7.1 understanding and learning of the knowledge of the visual objects is presented, in Section 7.2 understanding and learning of the knowledge of the sensory objects is presented and in Section 7.3 understanding and learning knowledge of the text objects is presented. Understanding and learning of the knowledge of the visual objects described in Section 7.1 is considered as acquiring the complex perceptual skills by SUS. Acquiring the complex perceptual skills by SUS, needed in perceiving the world, is connected with implementation of the sophisticated image processing algorithms and learning complex patterns of the visual reasoning sequences. The generalization and the specialization, the important forms of visual abstraction are described in Sections 7.1.1 and 7.1.2. The short description of learning of the visual knowledge called the categorical learning was presented in [1].

In Section 7.1.3. the new developed categorical learning techniques, learning by the small alternation of a visual object (LSA), learning from the 'simple to complex' (LSC), learning by simplification of the complex object (LSCO), learning from parts (LP), and learning parts decomposition (LPD), are described. The new developed categorical learning techniques, applied to learn knowledge of the selected ontological categories, is described in Chapter 8. Understanding and learning knowledge of the sensory objects, described in Section 7.2 is considered as acquiring by SUS the complex perceptual skills, whereas understanding and learning of the knowledge of the text object, described in Chapter 7.3, is considered as acquiring of the complex interpretational skills. Acquiring the complex interpretational skills by SUS, needed in finding of the meaning of the text, is connected with learning of the query-form, the basic-form, the procedural-form and the explanatory and interpratational script. In this book learning and understanding of the text that belongs to one of the text categories, such as the category of text-queries, the category of text-tasks, or the category of dictionary-texts, will be presented. The educational tests described in Chapter 7, that are used to test the ability of students to understand, are applied to test the understanding abilities of SUS. The knowledge implementation approach described in Chapter 7 is used for learning knowledge and skills of the different categories of objects.

Chapter 8 presents the knowledge implementation, understood as learning of the knowledge and skills. Learning of the knowledge and skills connected with understanding of the text will be called learning of the new knowledge, whereas learning of the knowledge and skills connected with understanding of the visual or sensory objects will be called learning of the new skills. In Section 8.1 learning of the new skills is presented, whereas in Section 8.2 learning of the new knowledge is presented. Learning of the new skills and new knowledge are processes that are strictly related. Learning of the different skills to deal with processing of the different types of information is approached by designing and implementing of the different algorithms. The implemented algorithms are usually applied to solve the very specific processing problems. The complex problem such as a visual reasoning is solved by implementing many different specific algorithms that are governed by a general algorithm of the visual reasoning. Learning of the new knowledge requires implementing sophisticated algorithms that make it possible processing learned knowledge. Also learned new skills, implemented in the form of the complex processing methods, are designed in the context of the specific properties of the learned knowledge. Learning of the new skills, presented in Section 8.1, is strictly connected with learning of the new knowledge. In the context of learning of the knowledge and skills of the visual objects and sensory objects, we will often refer to this problem as to learning of the new knowledge of the visual and sensory objects. Learning of the knowledge of the visual objects is connected with application of the complex image processing methods where learned knowledge is extracted from the imagery data at the perceptual level. This can be seen as learning of the specific perceptual skills to acquire data from the image.

The knowledge implementation is part of the shape understanding method that is aimed at building the understanding/thinking machine, the shape understanding system (SUS). The short description of the shape understanding system is presented in Section 8.1.1; however in this book problems connected with the implementation of the shape understanding system (SUS) will not be discussed. The reason is that the theoretical issues, connected with learning and understanding are very complex, and an attempt to describe the implementation problems could, instead of clarifying things, make them less understandable in the context of the material presented in this book.

In Section 8.1.2 learning of the knowledge of the visual object is presented. Learning of the knowledge of the visual object can be seen as learning of the specific perceptual skills to acquire data from the image. In Section 8.1.2.1 learning of the knowledge and skills of the specific categories derived from the sign category is presented. Knowledge of the specific categories derived from the sign category is learned at the prototype level. In Section 8.1.2.2 learning of the knowledge and skills of the letters category is presented. Learning of the knowledge of the letters category is focused on learning of the visual knowledge of letters from all existing writing systems and non-visual knowledge connected with using a letter during reading and understanding of the text. In Section 8.1.2.3 learning of the knowledge of the arrow category is presented. In Section 8.1.2.4 learning of the knowledge of the category of road signs is presented and in Section 8.1.2.5 learning of the knowledge of the knife category, as an example of learning of the knowledge of the real world objects category, is presented. Learning of the knowledge of the category of sensory objects is presented in Section 8.1.3. Learning of the procedural-form, presented in Section 8.1.4, is connected with learning of problem solving skills that are used to solve educational tasks. Learning of new problem solving skills is concerned with learning of the new methods of solving the problem and implementing of new methods of processing and storing the acquired knowledge. Learning of problem solving skills involves learning of the knowledge and mechanisms how to apply acquired knowledge to obtain the solution of a problem (a text-task).

In Section 8.2, learning of the new knowledge is presented. As it was described, learning of the new skills and new knowledge are processes that are strictly connected. Learning of the knowledge of the visual objects can be seen as learning of the specific perceptual skills to acquire data from an image and because of this, learning of the knowledge of the visual object is presented in Section 8.1 as learning of the new skills. Learning of the new knowledge that is used during understanding of the text requires transfering of the knowledge from the different available sources and means, such as dictionaries or handbooks, into SUS. In this book, learning and understanding of the text that belongs to one of the text categories such as the category of text-query, the category of text-task or the category of dictionary-text, is presented. Knowledge associated with the text categories includes the categorical chain, the coding category, the query-form, the basic-form, the procedural-form, the explanatory script and the interpretational script. The categories that are basic components of acquired knowledge by SUS are presented in Chapter 8.2.1. Learning of the categories of the visual objects represented by the categorical chain is described in Section 8.2.1.1. Learning of the knowledge categories is described in Section 8.2.1.2. Learning of the coding categories is described in Section 8.2.1.3. Learning of the knowledge schema is described in

Section 8.2.2. Learning of the knowledge of the text-query is described in Section 8.2.3. Learning of the knowledge of the text-task is described in Section 8.2.4. Learning of the knowledge of the query-form is described in Section 8.2.4.1. The query-form is the generalization of the text-task expressed in the form of the clauses consisting of the coding categories. In Section 8.2.4.2 learning of the basic-form that refers to the basic meaning of the text-task is presented. In Section 8.2.4.3 learning of the explanatory script is presented. The mathematical-text-task, the physical-text-task, the geographical-text-task are the different categories of the educational tasks. In Section 8.2.4.4 learning of the mathematical-text-task is presented. In Section 8.2.4.5 learning of the physical-text-task is described. In Section 8.2.4.6 learning of the geographical-text-task is described. In Section 8.2.4.7 learning of the tools-text-task is described. In Section 8.2.5 learning of the dictionary-text is described.

In this book, the new term knowledge implementation is introduced to denote the new method of the meaningful learning in the context of machine understanding. A proposed method of the knowledge implementation is aimed at the building of the machine, the shape understanding system (SUS) that has the ability to think and understand. The knowledge implementation is part of the shape understanding method [1], developed by authors that stresses the importance of the knowledge acquisition in understanding and thinking processes. The term knowledge implementation introduced in this book refers to learning of the knowledge and skills by the machine, the shape understanding system (SUS). The knowledge implementation is defined in the context of both human learning and machine learning. However, there is a considerable difference in acquiring knowledge and skills by SUS and acquiring knowledge and skills by a student (human being).

In this book the term machine learning, in contrast to the meaning of existing term machine learning, is defined as meaningful learning in the context of the newly defined concept of the knowledge implementation. In the context of the newly defined term machine learning as the meaningful machine learning, the new concept machine understanding or "SUS understanding" is introduced. Machine understanding is defined as the ability of the machine to understand of the perceived object or meaning of the text. The term machine understanding similarly like the term of machine learning needs to be regarded as the two different aspects of the knowledge implementation. However, machine understanding is different in comparison to our human understanding. Machine understanding can be used to explain some of the brain processes connected with understanding, however it will not be possible to explain all subtle processes that govern our human understanding by applying the computational model.

Learning of the new knowledge is connected with learning of the knowledge that is used during understanding of the text. Learning of the knowledge and skills connected with understanding of the text will be called learning of the new knowledge, whereas learning of the new knowledge and skills connected with understanding of the visual or sensory object will be called the learning of the new skills. The knowledge implementation is focused on learning of the knowledge and skills by a machine (SUS) and is concerned with two main aspects of SUS learning: learning of the visual knowledge in the context of the categorical structure of the learned categories of the visual objects and learning of the knowledge that is connected with understanding of the content of the text. Learning of the different skills to deal with processing of the different types of information is approached by designing and implementing of the different algorithms. The implemented algorithms are usually used to solve the very specific processing problems. The complex problem such as a visual reasoning is solved by implementing many different specific algorithms that are governed by the general algorithm of the visual reasoning. Learning new skills by SUS is approached by acquiring of the diverse problem solving skills and implementing of the different processing and reasoning methods. The important problem solving skills are skills used to solve educational tasks. SUS not only acquires the new knowledge but also develops new problem solving skills by learning new problem solving methods and implementing new methods of processing and reasoning. Learning of problem solving skills starts with the more basic problems and progresses to those that are most difficult. Most often many methods can be applied to solve a problem (text-task). The solution methods are usually related to each other and SUS needs to learn those methods to solve many different problems for each specific topic and to learn to identify similar problems. Problems are regarded as similar in the context of the similarity at the query-form level, the basic-form level or the procedural-form level.

The knowledge implementation is concerned with learning of the visual knowledge and non-visual knowledge. Proposed new learning methods, that are part of the knowledge implementation method, are applied to learn knowledge of objects from the selected categories such as the category of visual objects, the category of sensory objects or the category of text objects. In the knowledge implementation approach, learning of the new knowledge that comes from the experience of perceiving of the world is connected with acquiring knowledge about visual objects. Learning of the knowledge of the visual objects is considered as acquiring the complex perceptual skills by SUS. Acquiring the complex perceptual skills by SUS, needed in perceiving of the world, is connected with implementing of the sophisticated image processing algorithms and learning of the complex patterns of the visual resoning sequences. During learning of the knowledge of the visual objects at first, the representative sample of objects from the learned category is selected and then, for each object, the symbolic name is obtained. The visual knowledge is learned based on the sample of the visual objects that represent the category of visual objects selected for learning and utilization of the different forms of visual abstraction. Learning of the knowledge of the visual objects is focused on acquiring basic knowledge of the visual categories. Learning of the basic knowledge of the visual object is connected with learning of the visual concept, the categorical chain and the knowledge scheme.

In our approach (knowledge implementation), all new learned knowledge is connected with the categorical hierarchical structure of the previously learned knowledge. The categories of visual objects, the categories of the knowledge and the coding categories are used to transform learned knowledge into the SUS inner complex categorical hierarchical structure. Knowledge needed in naming of the perceived object is represented as the hierarchical structure of the categories of the visual objects. Knowledge of the visual object includes the visual concept that represents the visual appearance of the prototype of that category. The learned non-visual knowledge, that is needed to interpret a text, is represented by the query-form, the basic-form, the procedural-form, the explanatory script and the interpretational script. Learning knowledge of the sensory object is to learn the sensory concept (model), the categorical chain and the sensory knowledge scheme. Learning knowledge of the text object is to learn the coding category, the query form, the basic-form, the procedural-form, and the explanatory and interpretational scripts.

The SUS ability to understand will be called machine understanding or SUS understanding. The term machine understanding, similarly like the term machine learning, needs to be regarded as the two different aspects of the knowledge implementation. However, machine understanding is different in comparison to our human understanding. Machine understanding can be used to explain some of the brain processes connected with understanding, however it will not be possible to explain all subtle processes that govern our human understanding by applying the computational model. SUS learns to understand the objects that belong to the different categories. The SUS ability to understand is related to the different categories of objects such as the category of the visual objects, the category of sensory objects and the category of text objects. Understanding of the object, which belong to the category of visual objects, requires engaging different skills and knowledge. Understanding of the objects from the different categories of objects is always connected with learning and utilizing learned knowledge.

SUS acquires the new knowledge that comes from an experience in perceiving objects that belong to the category of visual objects, and from many different sources that are available within the framework of the school system. SUS learns knowledge that is part of the curriculum program applied at schools, for example, knowledge of subjects such as mathematics, physics, chemistry, biology or geography, following general syllabus, which specifies what topics must be understood, and to what level to achieve a particular grade or standard. The way in which SUS acquires the new knowledge is based on the learning schema, which specifies what topics must be understood, and to what level to achieve an assumed standard. The learning schema defines the way of learning of the new knowledge in the context of SUS understanding.

SUS learns incrementally via a cycle of learning the new knowledge and then testing the new knowledge in the context of all learned knowledge. Learning is an iterative process and learned knowledge is tested and corrected at each iterative stage. SUS learns during incremental learning process which is time consuming, however each new version of SUS contains improved learned knowledge. Each copy of a given version of SUS has the same ability to understand. Development of new versions of SUS is time consuming, however learned knowledge can be passed from one copy of SUS to another.

Knowledge learned by SUS is tested during the understanding process. SUS is tested in the context of efficiency of learning of new skills as well as in the context of the ability to understand objects from the visual category and the ability to solve problems from the selected topics from subject such as mathematics or physics. In the context of the knowledge implementation, testing the ability of SUS to learn new skills and new knowledge is called SUS assessment. SUS assessment can be based on measurement of the SUS ability to solve educational tests that are applied for testing students' performance at the school. In order to evaluate the ability of SUS to understand the text-tasks, standardized student achievement tests can be applied. It is assumed that SUS is able to understand the text from the selected text category if SUS can solve different tests such as standardized student achievement tests. It is assumed that if SUS is able to respond correctly to questions from the educational tests then it possesses the ability to understand. In this context SUS understanding ability is related to the different scientific domains such as geography, physics or chemistry and the educational tests from those domains domains are applied to examine SUS ability to understand.

During SUS learning the main source of the knowledge of the selected subject is acquired from books that are used by students at school such as handbooks or books with suplementary materials. Learning to understand of the text requires transfering of the knowledge from the different available sources and means, such as dictionaries or handbooks into SUS. Similarly to a student who learns to pass a test, SUS learns different texts, from different achievemnt tests, that belong to the different text categories in order to "perform well" on a student achievment test. The student achievement test is a test used for monitoring students' perforamnce at school. These types of tests are used to assess the ability of SUS to understand learned knowledge.

Following the way of learning at a school, SUS learning is controlled by the difficulty factors, that means, at first SUS learns to comprehend simple texts and solve simple problems that are suitable for a student from the primary school. In order to understand the complex problem from a given domain SUS needs to be able to understand the basic problems from that domain. Similarly like a student, SUS will have difficulty in solving problems in the areas where SUS does not learn proper skills and knowledge. In order to improve the results of learning in these areas, SUS needs to learn to solve many prolems that will cover material that a student is learning at school. In the case of an error during learning process the error can be easy to correct during the testing stage. A complete step-by-step explanation for the solution of a problem is very useful during SUS learning. Learned step-by-step explanation for the solution makes it possible to acquire the ability to provide the explanation for solved problems. It should be noted that to understand and to solve a problem does not mean to be able to provide an explanation for solved problem.

SUS operates in two main modes, the learning and understanding mode. Learning and understanding are complementary processes. SUS ability to understand depends on the effectiveness of learning process and learning of the new knowledge depends on the ability of SUS to understand. The knowledge implementation is based on an assumption that a system to be able to understand needs to acquire the knowledge and skills and acquired knowledge needs to be fully understood. Understanding of the learned knowledge means that there are meaningful relations among learned facts stored in memory. Understanding of the text is associated with learning both knowledge and skills. Understanding of any arbitrary text is a very difficult task. The text can be written in any possible language and it can have very different meanings. Understanding of the educational text is related to the different scientific domains such as mathematics, physics, geography or chemistry. Finding of the meaning of the text requires learning the vast amount of knowledge. Similarly to a student, SUS will have difficulty in understanding of the text in the area where SUS did not learn skills and knowledge. The educational tests used to test students' understanding abilities are applied to test the understanding ability of SUS. Understanding of the text requires well developed problem solving skills. The most important skills in problem solving are skills in solving educational tasks. Solving educational tasks is performed in two main stages, in the first stage SUS understands the task and in the second stage SUS solves the task. Understanding of the educational task is considered at the two different levels. At the first level, the type of problem is identified and solved and, at the second level, the real world situation to which the problem referred to is reconstructed.

Understanding of the object from the different categories of objects is always connected with learning and utilizing learned knowledge. During understanding, that depends on acquired knowledge and skills, the perceived object is linked through its name with all previously acquired knowledge. SUS ability to understand is related to the different categories of objects such as the category of visual objects, the category of sensory objects or the category of text objects. Understanding of the object, which belongs to the category of visual objects, requires engaging different skills and knowledge.

The very important role in the understanding of the visual or sensory object plays the correct naming of the object whereas finding the meaning of the text plays a key role in the understanding of text objects. When an object from the category of text objects is understood by finding of the meaning of the text, an object from the category of visual objects or sensory objects is understood by naming of the examined object. The visual object, after naming, is interpreted based on knowledge of the ontological visual category and knowledge of the knowledge scheme. Understanding of the complex visual object is to know from which parts it consists of, what material is it made of, what is it used for, how to make it, how to use it and how does it work.

Problems connected with understanding are very complex and it is not possible to apply one general method that can solve very specific problems at the different levels of understanding. In this book, many new terms and concepts are introduced in order to describe and explain some issues connected with building of the understanding machine SUS. These terms are explained by referring to the content of our book and other our works rather than to existing literature in related areas of research. In order to describe so diverse and complex problems that are connected with building the understanding system, there is a need to create the theoretical framework. The theoretical framework makes it possible to keep the explanatory description understandable in the context of the results from the previous research on the development of SUS.

This book raises many questions that are discussed in the area of cognitive science, linguistics and philosophy of mind. In this book, these questions will be only very briefly described in the context of the newly introduced concept of the knowledge implementation. In the next our book the problems that are only touched in this book will be discussed in more details.

The possibility of building of the intelligent machines became reality in the time when the first computer was built. The performance of designed intelligent systems is compared to performance of the human being. In addition, application of more and more powerful intelligent systems has influence on many different areas of human activity and human thinking. In this context, the question "What is the basic difference between today's computer and an intelligent being?" is often formulated by contemporary philosophers. Many contemporary philosophers of mind believe, that the mind is literally a complex piece of computer software implemented on the hardware of the brain (see [2], [3]). This view is very popular among members of the scientific community as it would be a scientific truth discovered based on the results of the scientific investigations.

In this book, the term knowledge implementation is defined in the context of both human learning and machine learning. The knowledge implementation is the term that is introduced to stress the big difference between human learning and learning by a machine. Machine ability to learn is often used as an argument to support the materialistic view of mental processes. In the light of our research this argument can not support the claim of materialistic view of mental processes. In order to avoid mistakes caused by application of simply analogy based on utilization similarity of meanings of concept, such as learning or understanding, to justify the materialistic view of mental processes, new concepts that refer to the machine processes need to be introduced. The concept such as the knowledge implementation, that is used instead of the concept such as machine learning, makes it possible to avoid the meaningful associations that can lead to see our mental process through the reductionist filter.

SUS learns knowledge that is part of the knowledge learned by students within the framework of the curriculum program, which specifies what topics must be understood and to what level to achieve a particular grade or standard. In many countries national curriculum program is developed by an independent authority and approved by education ministers. The way in which SUS learns reveals the tendency of the existing educational institutions in supplying the learning environment that is more suitable for machine (robot) than for a human being. Also the latest research (see [4]), connected with introduction of testing support this general hypothesis. Introduction of testing encourages methods of teaching that promote shallow and superficial learning rather than deep conceptual understanding and the kinds of complex knowledge and skills needed in modern, information-based societies. As Alexander claims [5], nothing that literacy and numeracy skills are tested within NAPLAN without reference to the subject context in which those skills need to be applied. There is tendency that teachers have increasingly been "teaching to the test" and in this view that teacher increasingly becoming technicians, obliged to deliver a prescribed and narrow product into which they have had little input. According to Dawson educational system of Western culture becomes increasingly technological and pragmatic, undermining the longstanding emphasis on liberal learning and spiritual reflection that were hallmarks of the Christian humanism that created it. Dawson argues that Western civilization can only be saved by redirecting its entire educational system from its increasing vocationalism and specialization. The European Christian culture of the past saw theology as the queen of the science. Christian culture has eternity before it and for that reason a Christian culture is potentially far wider and more Catholic than a secular one. It is God centered, not man centered, and it consequently changes the whole pattern of human life by setting it in a new perspective [6].

We are aware that this book could be written in a different way where some issues could be explained in more details or presented in the different way. We would like to explain that this book was written in "special" conditions. During the most crucial part of writing of this book, we were notoriously expelled from our own flat where most of work connected with preparation of this book was carried out.

Contents

1 Learning

The knowledge implementation is the term that denotes acquiring knowledge by a machine stressing the big difference between human learning and learning by a machine. The knowledge implementation is concerned with two main aspects of human learning, learning knowledge and learning skills, and is defined in the context of both human learning and machine learning. In this Chapter some aspects of human learning, that are related to the newly introduced concept of the knowledge implementation, are described. In Chapter 2, short survey of the existing machine learning methods is presented.

This Chapter is not intended as a survey of literature on the vast topic concerning human learning, but rather a presentation of some aspects of learning considered to have implications for material presented in other Chapters. This book presents the latest selected results of research on the further development of the shape understanding method the aim of which is to build a thinking machine. The shape understanding method was described in the previous book of authors titled "Shape Understanding System: the first steps toward the visual thinking machines" [1].

Learning, which is the process of acquiring knowledge, is defined as "a process of building and refining a knowledge structure towards a form which is optimally efficient for the several functions it must serve" [7]. Learning is connected with acquiring new or modifying existing knowledge, behaviors, skills, values, or preferences. In this book, we will focus on two aspects of learning, connected with acquiring of the new knowledge or modifying the existing knowledge and skills. Learning is a social process that depends on the cultural development of the society and is strictly connected with a concept formation and language development processes. During learning the new knowledge is acquired and new skills are formed. Human learning may occur as part of education, personal development, schooling or training. Learning may be goal-oriented and may be aided by motivation.

One of the main goals in pursuing of new knowledge is to understand the world in its material and cultural aspects. The capacity to categorize enables humans to handle everyday information. Once categories have been learned they can be used without further learning. The learning and usage categories are elementary forms of cognition by which humans adjust to their envinronment. A key element of the knowledge that is stored in our brain is the concept that is often called a 'category'. A concept is mental activity that brings together two or more situations, experience or objects into relationships. Via concept , past experience can be used in a given situation. As concept evolve progresively with experience, a person's ability to understand and deal with some domain may improve as well [8]. Gennari at al. ([9]

Z. Les & M. Les: Shape Understanding System - Knowledge Implementation and Learning, SCI 425, pp. 1–7.
springerlink.com

view human learning as a gradual process of concept formation. According to Gennari, a hierarchy of concepts that summarize and organize the experience, is the result of the learning process. The concept was often viewed in relation to the universal terms. In the Middle Ages the problem of universal terms or class names was the topic of many tractates [10]. These universal terms were thought of as a hierarchical structure of the class names. Categories are topic of research in philosophy, psychology and linguistics. In linguistics research [11] categories are viewed as logical bounded entities, membership in which is defined by an item's possession of a simple set of critical features. It is claimed that some natural categories are analog and must be represented logically in a manner which reflects their analog structure. They are internally structured into the prototype and non-prototype members.

As it was described, during learning new knowledge is acquired and new cognitive skills are formed. Human acquisition of cognitive skills is explained by ACT theory developed by Anderson [12]. According to Anderson all higher-level cognitive functions are achived by the same underlying structure, that means the same data structures and processes can be used in programs for language and for problem-solving. Anderson argues that may be a significant innate component to succesful of cognitive skills, however he claims that it is implausible for human to evolve special faculties for mathematics or physics.

In literature, many different types of learning are described. The simplest is an animal learning, based on habituation that is non-associative learning. An animal first responds to a stimulus, but if it is neither rewarding nor harmful, the animal reduces subsequent responses.

Meaningful learning is based on the assumption that learned knowledge is fully understood and that there is a relation among specific facts stored in memory. In contrast to rote learning, that is based on memorizing learned knowledge, meaningful learning is based on understanding of learned knowledge. Rote learning avoids understanding the inner complexities and inferences of the subject that is being learned. The memorized material can be recalled by the learner exactly the way it was read or heard. Rote learning, in which a big role plays memorizing of learned material rather than understanding of the learned knowledge, is important in learning for passing assessment tests. The discrimination learning is defined as the process by which a subject learns to make different responses to different stimuli. Discrimination learning, where a subject learns to respond to a limited range of sensory characteristics, is important part of understanding process. It covers a diverse range of different types of learning, such as perceptual learning, concept learning and language learning.

Learning through the individual experience is connected with learning of the knowledge of the visual objects. Child learns by perceiving and manipulating of an object. Acording to Piaget [13] learning knowledge about objects starts in the early childhood (infancy) and is based on interactions and mobility that allows child to learn new things. In that stage of child development intelligence is present, motor activity and limited knowledge is developed, and some language skills are developed at the end of this stage.

Associative learning is based on an assumption that ideas and experiences reinforce one another and can be linked to enhance the learning process. Associative learning is based on the ability to connect a previously irrelevant stimulus with a particular response. It occurs through the process of conditioning, where reinforcement establishes new behavior patterns. Essence of association lies in an observation that a subject perceives something in the environment (sensations) that result in an awareness of idea of what is perceived. Associations are based on similarity, frequency, salience, attractiveness and closeness of objects or events in space or time. In the context of learning knowledge of the visual object that is part of the research on development of the thinking machine the perceptual learning can be used to explain some processes relevant to the learning process. The perceptual learning [14] is governed by "stimulus equivalence" or "stimulus generalization." Perception in the broader sense must include mental imagery and its relation to direct sensory observation. Influence of memory on the perception needs to be taken into account in explanation of the visual perception. A perceptual act is never isolated and is modified by the learned or memorized facts that were perceived in the past. In order to transfer the visual perceptual material into the categories that are learned in the past, there is a need for visual concept that can combine the visual and non-visual data. Visual knowledge acquired in the past helps not only in detecting the nature of an object but it also assigns the present object a place in the system of things constituting our total view of the world. Visual knowledge that is stored in memory needs to supply the working hypotheses, called expectation, about the possible perceptual object. Visual knowledge and correct expectation facilitate perception whereas inappropriate visual concepts will delay or impede it. The percept must define the object clearly and must resemble sufficiently the memory image of the appropriate category. Often, however, there is enough ambiguity in the stimulus to let the observer find different shape patterns in it as he searches for the best fitting model among the ones emerging from the memory storage.

The inference theory, associated with the empiricist view, argues that knowledge is acquired solely by sensory experience and association ideas. The mind at birth is a tabula rasa upon which experience records sensations. Helmholtz postulated the existence of "primary" percepts and claims that the primary percept contains all the distortions of projection but judgment intervenes and corrects them. Helmholtz assumed that these corrections are based mainly on knowledge previously acquired and he later argued that we learn to interpret percepts through a process of association. The Gibson theory [15] claims that sensory input is enough to explain our perceptions and he points out that the Gestalt school has been occupied with the study of two-dimensional projections of the three-dimensional world and that its dynamism is no more than the ambiguity of the interpretation of projected images. Marr [16] made significant contributions to the study of the human visual perception system as corresponding to modules of the human visual system. Gestalt psychologists believe that important learning processes involve a restructuring of relationships in the environment, not simply associating them. Psycholinguists indicate that some of the aspects of language learning such as a native "grammar" can be inherited genetically. The contemporary theories of learning indicate on the role of motivation in learning process. Scientists are looking for universal principles governed all learning processes that could explain the mechanism of learning process. Rigorous, "objective" methodology was attempted so that the behavior of all organisms could be comprehended under a unified system of laws modelled on those postulated in the physical sciences. However, today most psychologists believe that no single theory of learning may be appropriate. The last attempts to integrate all knowledge of psychology into a single, grand theory occurred in the 1930s. Three thinkers Guthrie, Hull and Tolman had a big impact on the theory of learning. Guthrie [17] claimed that responses (not perceptions or mental states) were the ultimate and most important building blocks of learning. Hull [18] argued that "habit strength," a result of practiced, stimulus-response (S-R) activities promoted by reward, was the essential aspect of learning, which he viewed as a gradual process. According to Tolman [19] learning is a process that is inferred from behaviour.

The cognitive learning is defined as a process for utilizing experimental inputs to shift representation so that an increasing variety of problems can be recognized, formulated, or coped with [20]. According to cognitive learning activity such as recall, calculate, discuss, analyze, problem solve are most important in learning of the conceptual knowledge. The psychomotor or affective ability are used during learning to dance, swim, or to like something or someone, love. These domains are not mutually exclusive. For example, in learning to play chess, learning the rules of the game (cognitive domain) is learned in parallel with learning how to set up the chess pieces on the chessboard and also how to properly hold and move a chess piece (psychomotor).

Active learning occurs when a person takes control of their learning experience. Since understanding of the information is the key aspect of learning, it is important to recognize what is understood. For example, a student can monitor his progress in understanding of the learned material. Active learning is focused on verbalizing of understanding of the learned parts. Informal learning occurs through the experience of day-to-day situations such as a play, or an exploring, whereas formal learning is learning that takes place within a teacher-student relationship, such as in a school system. Nonformal learning is organized learning outside the formal learning system e.g. in clubs organizations, workshops.

Formal learning is learning that takes place in a school system. Learning through the educational process is connected with schooling from the primary to the higher education level. Education is the formal process by which society transmits its accumulated knowledge, skills, customs and values from one generation to another. The education is the object of research of the sociology, the psychology and existing education theories tried to explain the processes connected with teaching and learning at school. Education theory refers to either a normative or a descriptive theory of education. A theory provides the goals, norms, and standards for conducting the process of education and hypotheses that have been verified by observation and experiment. A descriptive theory of education can be thought of as a conceptual scheme that ties together various discrete particulars. For example, a cultural theory of education shows how the concept of culture can be used to organize and unify the variety of facts about how and what people learn. Educational psychology is concerned with the study of how humans learn in educational settings, the effectiveness of educational interventions, the psychology of teaching, and the social psychology of schools as organizations. Educational psychology can in part be understood through its relationship with other disciplines.

Learning at a school is based on a curriculum program. The curriculum is prescriptive, and is based on a more general syllabus, which specifies what topics must be understood, and to what level to achieve a particular grade or standard. Recently, there is a tendency in many countries to replace a state curriculum with a national curriculum. The national curriculum is developed by an independent authority and approved by education ministers. Schools and states are responsible for implementation of the national curriculum. Each state is developing its own implementation plans. The curriculum is a set of courses, and their content, offered at a school or university. The curriculum program describes the body of the knowledge that is learned by students. In the country such as Australia education follows the three-tier model which includes primary education (primary schools), followed by secondary education (secondary schools/high schools) and tertiary education (universities and/or TAFE Colleges). The education system in Australia consists of a total of 12 years. Primary schools and high schools are based on the age of the student, so that every room has the same age group, with a student hardly having to repeat a year. The students with problems in their studies may be put into special classes to help with academic deficits. In general, primary education consists of six or eight years of schooling starting at the age of five or six, although this varies between, and sometimes within, countries. The division between primary and secondary education is somewhat arbitrary, but it generally occurs at about eleven or twelve years of age. Depending on the system, schools for this period, or a part of it, may be called secondary or high schools, gymnasiums, lyceums, middle schools, colleges, or vocational schools. Higher education, also called tertiary, third stage, or post secondary education, is the non-compulsory educational level that follows the completion of a school providing a secondary education, such as a high school or secondary school. Tertiary education is normally taken to include undergraduate and postgraduate education, as well as vocational education and training. Colleges and universities are the main institutions that provide tertiary education. Tertiary education generally results in the receipt of certificates, diplomas, or academic degrees. University education includes teaching, research and social services activities, and it includes both the undergraduate level and the graduate (or postgraduate) level. In Australia there are Government schools, Catholic schools and Independent schools. Government schools educate approximately 65% of Australian students, with approximately 34% in Catholic and Independent schools. Regardless of whether a school is part of the Government, Catholic or Independent systems, they are required to adhere to the same curriculum frameworks of their state or territory. The curriculum framework however provides for some flexibility in the syllabus, so that subjects such as religious education can be taught.

Learning needs to be evaluated in order to find if the learning achived its goal. Evaluation of the learning progress is performed during the educational assesment.

The educational assessment is the process of documenting, usually in measurable terms, knowledge, skills, attitudes and beliefs. Assessment can focus on the individual learner, a group of learners (class, school) or the educational system as a whole. The important part of the educational assessment is testing of the students' achievements as well as testing general ability of students. In the context of this book however, of interest will be assessment of learning outcomes of the individual student in terms of his achievements in a subject domain. A good assessment has both validity and reliability. A valid assessment is one which measures what it is intended to measure and a reliable assessment is one which consistently achieves the same results with the same (or similar) cohort of students. In order to monitor the results of learning at both the individual and the school level the different methods of testing are used. There are tests for each learned subject such as mathematics, physics, chemistry, biology, geography, history or English. Tests can offer useful information about student progress and curriculum implementation, as well as formative uses for learners. The test is developed by the test development team consisted of staff from the test development sections. The process of test development includes: writing and reviewing items, pre-piloting the test material, preparing marking guides and marker training material, and selecting items for the main study. The tests consist of number of questions on selected topic. When administered in useful ways, tests can offer useful information about student progress and curriculum implementation, as well as offering formative uses for learners. Standardized testing nationwide aligns with state curriculum and link teacher, student, district, and state accountability to the results of these tests. The important part of the educational assessment is testing of the students' achievements as well as testing general ability of students [21]. Testing of the students' achievements is carried out in the framework of monitoring of the school performance and in Australia is carried out within the National Assessment Program: Literacy and Numeracy (NAPLAN) [22]. Those educational assessment programs are a curriculum-based assessment for Year 3, Year 5, Year 7 and Year 9 students, testing knowledge and skills in Literacy and Numeracy. The results of the tests provide information for students, parents, teachers and principals about student achievement, which can be used to support teaching and learning programs.

The discipline of developing educational tests and measurement procedures is referred to as psychometrics. The central questions are of reliability and validity of test scores. Item response theory is currently the most often applied psychometric theory to estimate the true score of the subject. In Australia the measurement of the test outcomes is based on the Rash model implemented as the Conquest program. The result of the measurement by applying Conquest describe a characterisitc of a subject as a numerical score. Review of the latest literature on this topic can be found in [4].

In order to facilitate learning and help to improve the learned results there are many different educational resources such as books or computer programs [23]. Knowledge is transferred to student by applying the different learning/teaching methods. Knowledge is acquired by using the different educational means but the most important source of learning material are books. Books, such as handbooks usually cover the material that is subject of learning at primary, secondary, or teritiary educational level. The books used at primary level refers to the simple knowledge. Example of series books used at the primary level is Excel basic Skills Homework [24] books aim of which is to build and reinforce basic skills in reading, comprehension and maths. The series support schoolwork by maintaining skills, therefore allowing students to learn new concepts always needing to revise past work.The series has seven core books , one each for years 1 to 7. For example, a book that refers to the student skills for student year 2 age 7-8 has carefully graded units. Each unit has work on numbers, measurement and shapes in math and comprehension, grammar, punctuation, spelling and vocabulary in English. If student finds he continually has difficulty in certain area, then there is a Basic Skills book that will take the student through the topic step by step. Each unit consist of questions on selected topic. The topic include numbers, measurement and shape in maths and comprehension, grammar, punctuation, spelling and vocabulary in English.

Many books that cover material that is subject of learnig at secondary, or teritiary educational level such as Schaum's outlines series offer solved and supplementary problems that are arranged to allow a progression within each topic. There are books for each learned subject such as mathematics, physics, chemistry, biology, geography, history or English. These books consist of problems often called questions, items or tasks. Some books offer one model problem and many problems following the model to practice the skills in solving a given problem. For example, books such as "501 Algebra questions" [25] provide problems to build mathematical and algebraic skills and can be used to supplement instructions in mathematics classes. The book coveres many algebraic topics. The structure of the book follows a common sequence of concepts introduced in basic algebra. Books of this type usually offer solved and supplementary problems that are arranged to allow a progression within each topic.

2 Machine Learning

In the previous Chapter some aspects of learning that are related to the new introduced concept of the knowledge implementation were described. In this Section a short survey of literature on the vast topic concerning learning by machine is presented. Not only humans, animals but also some machines have the ability to learn. The complete and autonomous learning systems that start with the general inference rules and learning techniques, and gradually acquire more complex skills and knowledge through continuous interaction with the information-rich external environment, are designed to be applied in areas such as robotics. These systems utilize the sensory data such as sound and image and combine them during reasoning process. The research in robotics is focused on learning which makes it possible for a robot to move and perform required work ([26], [27], [28], [29], [30], [31]).

Learning knowledge by machine is often called knowledge acquisition. Knowledge acquisition is the process of acquiring knowledge from one or more sources and passing it on in a suitable form to human or machine. This process usually involves learning, reformalizing, transferring and representing the knowledge. According to definition presented in ([32]), knowledge acquisition is the transfer and transformation of problem-solving expertise from some knowledge sources to a program. The process of the knowledge acquisition involves the definition, implementation and refinement. It also involves choosing the knowledge representation to embody the facts and relations acquired from sources of the knowledge such as textbooks and other documented knowledge or human experts and their experience in the problem domain. According to Buchanan at al. [32] knowledge acquisition consists of following stages: identification, conceptualization, formalization, implementation and testing. Modern knowledge-based systems, called also expert systems, learn knowledge that is extracted from human experts that are specialists in a narrow domain. In the process of the knowledge acquisition, a knowledge engineer communicates with human experts in order to acquire the relevant knowledge. The knowledge engineer uses also books, manuals, case studies, and other material in order to understand the problem domain. Experts often have difficulties in describing their knowledge in the precise, complete and consistent way that is enough for use in a computer program. They sometimes give explanations for their decisions that do not correspond to their actual, perhaps unconscious or compiled, reasons for making the decisions [33]. This difficulty stems from the inherent nature of the knowledge that constitute human expertise ([32]).

Knowledge-based systems consist of two major components: the inference engine and the knowledge base. The inference engine contains the reasoning or

Z. Les & M. Les: Shape Understanding System - Knowledge Implementation and Learning, SCI 425, pp. 9–15.
springerlink.com

processing mechanism used by the system, that is, it contains the general problem-solving knowledge. The knowledge base contains the actual knowledge of a problem domain in some suitable format; that is, it contains the domain specific knowledge. MYCIN [34] or DENDRAL are well known examples of expert systems that pave the way for the new research in building the sophisticated knowledge based systems. The neural network based systems that learn knowledge from data, not only are applied to solve complex problems but also supply new facts about complexity of the brain processes. In robotics, the aim of which is to build a machine that will be able to act in a similar way as human being, the most of research is concentrated on a problem of navigation in unknown environment or to build a humanoid robot. However, the humanoid robot should not only look similar to the human shape but also it needs to possess the capabilities to understand to be able to exist in the human world.

Problems with manual knowledge acquisition have stimulated considerable research effort directed toward the development of an automated acquisition method. The automated knowledge acquisition is also called machine learning. Machine learning, a branch of artificial intelligence, is a scientific discipline concerned with the design and development of algorithms that allow computers to evolve behaviors based on empirical data, such as from sensor data or databases. A machine learning task is to construct the complete and autonomous learning systems that start with the general inference rules and learning techniques, and gradually acquire more complex skills and knowledge through continuous interaction with the information-rich external environment. Machine learning is concerned with the development of algorithms allowing the machine to learn via inductive inference based on observing data that represents incomplete information about statistical phenomenon, generalize it to rules, and make predictions on missing attributes or future data. An important task of machine learning is classification, which is also referred to as pattern recognition, in which machines "learn" to automatically recognize complex patterns, to distinguish between exemplars based on their different patterns, and to make intelligent predictions on their classes. Pattern analysis is an area of research that often deals with the recognition of a known object in the image. The research fields, such as content-based retrieval that locates relevant images in a large collection of images [35], [36], analysis of the symbols in a printed document [37], recognition of the script words [38], recognition of the handwritten letters [39], [40], [41], the face recognition [42] or the object recognition [43], are limited to perform recognition based on application of classical pattern recognition tools. This approach can give good results in the case when a number of objects that need to be recognized is rather small.

There are many research areas focused on learning knowledge from the data, for example machine learning, data mining or knowledge discovery. Machine learning focuses on the prediction, based on known properties learned from the training data, whereas data mining and knowledge discovery in databases focus on the discovery of previously unknown properties on the data. However, these two areas overlap in many ways: data mining uses many machine learning methods, but often with a slightly different goal in mind. On the other hand, machine learning also employs data mining methods as "unsupervised learning" or as a preprocessing step to improve learner accuracy. Machine learning algorithms can be classified into two major groups: supervised learning and unsupervised learning. The supervised learning generates a function that maps inputs to desired outputs called labels that are provided by human experts by labeling the training examples.

As it was described, machine learning is an area of research that is to investigate the possibility of automated knowledge acquisition by a machine. The different learning strategies and methods were proposed over the recent years. The best known are the decision tree ID3 [44], CART [45], STAR methodologies [46], Explanation-Based Learning [47], or the connectionist model [48]. However, there is no method that can be used to learn the visual knowledge.

There are various methods which are used in symbolic machine learning: symbolic empirical learning, explanation-based learning, case-based reasoning, and the integrated learning methods [49], [50]. Based on the learning strategy the following learning methods can be distinguished: learning by induction (learning from examples, learning from observation, learning by discovery), learning by deduction and learning by analogy. For example, inductive learning is an inductive inference from facts provided by a teacher or the environment. The process of inductive learning can be viewed as a search for plausible general descriptions (inductive assertions) that explain the given input data and are useful for predicting new data. These assertions form a set of descriptions partially ordered by the relation of relative generality. One of the types of inductive learning is conceptual learning from examples (concept acquisition), whose task is to induce general descriptions of concepts from specific instances of these concepts. An important variant of concept learning from examples is the incremental concept refinement, where the input information includes, in addition to the training examples, previously learned hypotheses, or human-provided initial hypotheses that may be partially incorrect or incomplete. In concept acquisition, the observational statements are characterizations of some objects pre-classified by a teacher into one or more classes. The induced hypothesis can be viewed as a concept recognition rule, such that if an object satisfies this rule, then it represents the given concept. Neural networks (a connectionist model) are often used in process of the knowledge acquisition. One of the most important features of neural networks is that they perform a large number of numerical operations in parallel. The distributed neural processing is typically performed within the entire array composed of neurons and weights [51].

A machine to be able to understand and think needs to have some mechanism that makes it possible to utilize knowledge during thinking process. In order to solve complex problems one needs both an appropriate knowledge representation and some mechanisms for manipulating that knowledge. Existing knowledge based systems apply the different methods of the knowledge representation. The knowledge representation is some chosen formalism for "things" we want to represent. There are two main important dimensions along which they can be characterized. At one extreme are purely syntactic systems, in which no concern is given to meaning of the knowledge. Such systems have simple, uniform rules for manipulating the representation. At the other extreme are purely semantic systems, in which there is no unified form. We can distinguish structures in which knowledge can be represented: production rules, semantic nets, frames, conceptual dependency and scripts. The production rules belong to syntactic systems because they usually use only syntactic information to decide which rule to fire. Semantic nets are designed to capture semantic relationships among entities, and they are employed with a set of inference rules. Semantic networks offer a convenient mechanism to describe semantics, syntax and pragmatics in the study of language [52]. The use of network structures is not new in knowledge representation. There are two major types of networks that deal with imprecise information and thus perform reasoning under uncertainty: Bayesian [53] and Markov [54]. Frame systems are typically more highly structured than are semantic nets, and they contain a large set of specialized inference rules. Conceptual dependence representation can be thought of as instances of semantic nets but having a more powerful inference mechanisms that exploit specific knowledge about what they contain [55]. Script (very similar to frames) in which slots are chosen to represent the information is useful during reasoning about a given situation. One of the methods of the knowledge representation that refers to the thinking process is representation that is based on the concept of frame. A frame is simply a data structure that consists of expectation for a given situation. A frame can consist of objects and facts about a situation, or procedures on what to do when a given situation is encountered. To each frame several kinds of information are attached. Some of this information is about how to use the frame, some is about what one expects to happen next, and some is about what to do if these expectations are not confirmed. Collection of frames is linked together into frame systems. The frame systems are linked, in turn, by an information retrieval network. A matching process tries to assign values to each frame's terminals which are partly controlled by information associated with the frame and partly by knowledge about the system's current goals. One of the advantages of this global model is that memory is not separate from the rest of thinking. The information-retrieval based on frame system explains differences between ways of thinking of the person, by assuming mechanism of quickly locating highly appropriate frames for "clever" persons. It indicates that good retrieval mechanism can be based only in part upon basic innate mechanism. It must also depend on (learned) knowledge about the structure of one's own knowledge. The short term memory is connected with sensory buffer and has also the suitable frames.

The neural networks approach uses "sub-symbolic computation" for description of the knowledge representation and its processing. The term "sub-symbolic computation" refers to the fact that, in distributed representations, a node is not associated with one particular symbol, being able to take part in the distributed representation of the various concepts. One of the most important features of neural networks is that they perform a large number of numerical operations in parallel. Almost all data stored in the network are involved in recall computation at any given time. The distributed neural processing is typically performed within the entire array composed of neurons and weights. Most of classical information processing models utilize symbolic sequential processing mechanism. In self-organization NN classes of objects are formulated on the basis of a measure of object similarity [56]. Most of measures of similarity are context free, that is, the similarity between any two objects A and B depends on the properties of the objects. In modified version of the BCM neuron, sets of these neurons, which are organized in lateral inhibition architecture, forces different neurons in the network to find different features based on similarity relation. It is similar to classical approach allowing, based on similarity of "concept", to reason by "analogy" in order to obtain knew description. However, neural systems seem to operate in a more 'holistic way' than inferential ones: they learn to associate entire input patterns with the corresponding output decisions. For example, the semantic maps are implemented based on the self-organizing feature map. The maps extract semantic relationship that exists within the set of language data-collection of words. The relationship can be reflected by their relative distances on the map containing words positioned according to their meaning or context. This indicates that the trained network can possibly detect the logical similarity between words from the statistics of the contexts in which they are used [51]. In more complex systems, for task as complex as the understanding of editorial texts or machine translations, mixture of the different ways of the knowledge representation and manipulation is utilized. For instance, in understanding editorial text, system abstract knowledge is organized by memory structures called Argument Units, which represent patterns of support and relationships among beliefs. When combined with domain specific knowledge, it can be used to argue about issues involving plans, goals, and beliefs in particular domain [57]. The hybrid method which is based on the knowledge graph and in which abstractions of the information and classification part of examples are explicitly stored is used for deriving production rules and generalization. These examples taken from field of AI can show how representation of the knowledge and its manipulation influence a way of perceiving some of the processes which have similar functions in the brain.

AI approach for knowledge representation gave only little attention to "pictorial knowledge" or tried to simplify pictorial information by use of "pictorial attributes". The psychology of language describes how semantic representations of utterances are elaborated. On a certain level of analysis, theories of meaning need not account for the processes which enable meaning to be expressed in mental representation derived from sensory modalities. Accounting for meaning calls for the coordination of abstract symbols through application of appropriate rules. Images can reflect even divergent semantic content but are identifiable with meaning of the sentence. Coding of a picture and coding of words have similar functional properties. The differences that subsist between the processing of picture and symbols are not assumed to reflect the existence of two distinct representational systems, each containing qualitatively different types of information. Differences between pictures and symbols stem from the different representation of image and symbol in memory. Pictorial information could be represented in memory in both modal and prepositional form [58].

Knowledge-based systems consist of the inference engine that contains the reasoning or processing mechanism. One area of research, which deals with process of reasoning, is artificial intelligence (AI). AI has shown great promise in application of the different forms of reasoning in the area of expert systems [59] or knowledge-based expert programs (see e.g. [60]) which, although powerful when answering questions within a specific domain, are nevertheless incapable of any type of adaptable, or truly intelligent, reasoning. In the field of the AI many forms of reasoning such as logic reasoning, statistical reasoning or fuzzy reasoning (see e.g. [61], [62], [63],) were employed. Logic reasoning forward or backward is often used in the logic programming (see e.g. [64], [65], [66]). One of the logic programming languages such as PROLOG applies the goal-directed reasoning (backward) in order to manipulate symbols (actually words) and find the solution. Another form of reasoning is statistical reasoning based on Bayesian statistics implemented in expert systems such as MYCIN [34] or Bayesian networks [67]. The distributed reasoning systems composed of a set of separate modules (often called agents) become one of the most popular (see e.g. [68]). Reasoning is often modeled as a process that draws conclusions by chaining together generalized rules, starting from scratch. In case-based reasoning (CBR) the primary knowledge source is not generalized rules but a memory of stored cases recording specific prior episodes. In CBR, new solutions are generated not by chaining, but by retrieving the most relevant cases from memory and adapting them to fit the new situation. CBR is based on remembering. Reminding facilitates human reasoning in many contexts and for many tasks, ranging from children's simple reasoning to expert decision making. Much of the original inspiration for the CBR approach comes from the role of reminding in humans reasoning. The quite extensive description of the different techniques of CBR can be found in [69].

There are systems that are built in order to interpret the perceived object or interpret an image, however these systems do not assume understanding of the perceived object or the real world scene. These systems are built based on the research in the area of computer vision and image understanding. The term image understanding has a range of meanings, but in general, image understanding refers to a computational, information processing approach to image interpretation. The term image understanding denotes an interdisciplinary research area that includes signal processing, statistical and syntactic pattern recognition, artificial intelligence, and psychology. Image understanding refers to knowledge-based interpretations of visual scenes that transform pictorial inputs into commonly understood descriptions or symbols (see for example: [70], [71], [72]). Computer vision is also used to refer to a similar research area ([73], [74]), but while computer vision emphasizes the computational aspects of visual information processing, such as measurement of three-dimensional shape information by visual sensors, image understanding stresses knowledge representation and reasoning methods for scene interpretation. Another field of research, that stresses modeling of the human visual system called computational vision, can also be treated as a field of image understanding research. Computational vision is a multidisciplinary and synergetic approach whose main task is to explain the processes of the human visual system and build artificial visual systems [75]. The aim of image understanding system often involves an identification of objects in images and the establishment of the relationships among objects. Image understanding systems use context, knowledge about the world, and machine learning techniques for knowledge acquisition. For example, a document image understanding system converts a raster image representation of a document into an appropriate symbolic form. It involves application of the research from many sub-disciplines of computer science that include image processing, pattern recognition, natural language processing, artificial intelligence and database systems. Image understanding systems interpret the image based on the knowledge that is stored in the form of the semantic networks. Examples of image understanding systems that utilize knowledge are VISIONS [76], ALVEN [77], ACRONYM [78], SCHEMA [79], SIGMA [71], CITE [80], SOO-PIN [81], or ERNEST [82]. There is also research in image understanding that is focused on interpretation of the medical images. The aim of this research is to interpret the image in terms of the diagnostic indicators [83]. When image understanding systems are focused on interpretation of the visual content of the image, they do not take into account such important issues as visual transformations, visual inference and all processes that are connected with communication in natural language of the results of understanding. The system that has abilities to understand the visual information needs to be able to solve tasks that are presented in the form of visual intelligence tests. Intelligence tests include tasks that deal with visual forms (shapes). In the present research, that is part of the shape understanding method, the shape is considered as a meaningful unit called a phantom. In literature the term 'shape' often refers to the geometry of an object's physical surface [16]. Existing methods of shape analysis are mostly concerned with shape recognition [84], [85] [86], [87], [88, 89], [90], [91], [92], [93] [94]. Visual systems applying shape as their knowledge are called model-based object recognition systems and have been used extensively by vision researchers [70], [95], [96], [97].

In the contrary to most of the machine learning methods our approach is based on assumption that knowledge needs to be learned not only from the data but also can be derived from knowledge that exists as the result of our human experience and learning. Also in parallel to learning knowledge the system needs to acquire new skills.

3 Learning – Machine Learning – Knowledge Implementation

In Chapter 1 some aspects of learning that is related to the new introduced concept of the knowledge implementation was described. In Chapter 2 a short survey of the existing machine learning methods was presented. In this Chapter, the knowledge implementation is defined in the context of both human learning and machine learning.

As it was described in Chapter 1, only humans, animals and some machines have the ability to learn. An animal learning is based on habituation that is non-associative learning. Existing machine learning systems start with the general inference rules and learning techniques, and gradually acquire more complex skills and knowledge through continuous interaction with the information-rich external environment. Most machine learning methods can be classified as some sort of rote learning. Rote learning described in Chapter 1, which is based on memorizing learned knowledge, is very different from meaningful learning that is based on understanding of learned knowledge. Meaningful learning is based on assumption that learned knowledge is fully understood and that there is a relation among specific facts stored in memory. SUS is the first attempt to build the system that has the ability to understand and, in the context of machine learning, the ability to understand by a machine can be called machine understanding. As it was described at the beginning of this Chapter, the knowledge implementation is defined in the context of both human learning and machine learning. However, machine learning is understood as meaningful learning in the context of machine understanding.

Learning knowledge by a machine was often called the knowledge acquisition. The knowledge acquisition is the process of learning knowledge from one or more sources and passing it on in a suitable form to human or a machine. In contrast to both animal and machine learning, human learning occurs as part of education, personal development, school or training. In the previous Chapter, the short survey of the different machine learning methods was presented. Although various machine learning methods were elaborated and many of them found application in building robots or expert systems, there is no method that can be used to learn knowledge offered by the curriculum program of the today's educational system. In the first Chapter, some aspects of human learning were described. Based on the analysis of human learning, especially in the context of learning at a school, the new method of learning by a machine was proposed by authors. This method, called the knowledge implementation, is concerned with two main aspects of human learning: learning of the visual knowledge in the context of the categorical

Z. Les & M. Les: Shape Understanding System - Knowledge Implementation and Learning, SCI 425, pp. 17–25.
springerlink.com

structure of the learned categories of the visual objects and learning of the knowledge that is connected with understanding of the content of the text. The knowledge implementation is focused on learning of the knowledge and skills by the machine (SUS). Similarly to a student engaged in learning process, the machine (SUS) can acquire the new knowledge and skills through learning process. However, there is a considerable difference in acquiring knowledge and skills by SUS and acquiring knowledge and skills by a student, so the term the knowledge implementation is introduced to stress the difference in acquiring knowledge and skills by SUS. Contrary to the most of machine learning methods that are focused on learning primarily from data, in our approach focus is not only on learning from data but also from knowledge cumulated as the result of scientific research and our human experience.

The knowledge implementation is based on assumption that a system to be able to understand needs to learn knowledge that is fully understood and that there is a relation among learned specific facts stored in memory. Human learning starts from learning by manipulating of the simple objects in the early childhood. However, the most important part of human learning is connected with formal learning that takes place within a teacher-student relationship in a school system. SUS is acquring new knowledge that comes from an experience in perceiving objects that belong to the category of visual objects and from many different sources that are avaible within the framework of the school system. Learning at a school is based on a syllabus which specifies what topics must be understood and to what level to achieve a particular grade or standard. The knowledge implementation, that defines the way in which SUS is acquring new knowledge, is based on the learning schema which specifies what topics must be understood and to what level to achieve an assumed standard. Learning schema defines the way of learning of the new knowledge in the context of understanding process. Learning of the new knowledge is to learn both visual and non-visual knowledge. Visual knowledge is connected with sensory knowledge and is learned as the knowledge of the visual and sensory objects. Non-visual knowledge is concerned with knowledge that is accesible through the written text. In further part of this book the term "SUS learning" will be often used as the synonym for the "knowledge implementation".

During learning, new knowledge is acquired and new skills are formed. Piaget's theory of developmental psychology [13] describes cognitive development from infancy to adulthood that is concerned with acquiring new knowledge and formation of new skills. Modern knowledge-based systems consist of two major components: the knowledge base and the inference engine that can be regarded as simple solving-problems skills. The knowledge base contains the actual knowledge of a problem domain in some suitable format. Similarly, in our approach, SUS learning is concerned with learning knowledge and learning skills. The knowledge base of knowledge-based systems is being incrementally developed via a cycle of testing and debugging. Isolating a specific bug may require a trace through dozens of inferences involving hundreds of facts. Similarly, SUS learns incrementally via a cycle of learning a new knowledge and testing new knowledge in the context of all learned knowledge. However, in contrast to the existing machine learning systems, SUS not only acquires new knowledge but also develops new problem solving skills by learning new problem-solving methods and implementing new methods of processing and reasoning. As it was described in Chapter one, ACT theory developed by Anderson [12] is the most widely applied theory of the human acquisition of cognitive skills that is a rule-based system in which rules are implemented in a psychologically and neurally manner. According to Anderson human mind is considered to be general-purpose, that is, the same data structures and processes can be used in programs for language and for problem-solving. Or in other words, all higher-level cognitive functions are achived by the same underlying structure. In our approach problem-solving skills are regarded as the complex processing-understanding mechanism where many different highly specialized algorithms are applied for processing information and performing reasoning. Implementation of the complex image processing algorithms, that deal with a very specific perceptual problem and application of the very specific reasoning methods coping with the transformation of the different tasks that are usually given as the text that needs to be understood, are part of the learned skills of the knowledge implementation method. In contrast to Anderson claims that human mind is considered to be general-purpose, the knowledge implementation approach is based on assumption that processing of the different information requires very specific processing methods. Implementation of these specific processing methods is understood as learning of the new specific problem-solving skills. These problems will be discussed in details in Chapter 8.

During learning problem-solving skills in a domain such as physics, the important part of learning physics is to acquire the ability to solve problems. Knowledge of physics consists of definitions, relationships, laws, rules and equations. Learning knowledge of physics means to learn physical facts, definitions, relationships, laws or rules. However, the ability to solve problems is the ultimate proof of understanding and competence in physics. Each learned subject such as mathemaics, physics or biology has its specific way of learning problem-solving skills. For example, learning physics is unlike learning most other disciplines. Physics has a special vocabulary that constitutes a language of its own, a language immediately transcribed into a symbolic form that is analyzed and extended with mathematical logic and precision. Words like energy, momentum, current, flux, inference, and so forth, have very special scientific meaning. These must be learned promptly and accurately because the discipline builds layer upon layer. Unless someone understands exactly what velocity is, he can not understand what acceleration or momentum are. Similarly in our aproach, SUS first learns knowledge from a domain such a physics by learning basic concepts from a selected topic such as "uniformly accelerated motion" or "work, energy, and power". Next, SUS learns to solve the simple problems from a selected topic and finaly the complex and difficult problems are learned.

As it was described in Chapter one, learning and usage of categories sometimes called concepts are elementary forms of cognition by which humans adjust to their envinronmentt. As concept evolve progresively with experience, a person's ability to understand and deal with some domain may improve as well [8]. According to Gennari [9] human learning is a gradual process of concept formation and a hierarchy of concepts that summarize and organize the experience is the result of the learning process. In our approach (the knowledge implementation), all new learned knowledge is connected with the categorical hierarchical structure of the previously learned knowledge. The categories of the visual objects, the categories of the knowledge and the coding categories are used to transform learned knowledge into the SUS inner complex categorical hierarchical structure. These problems will be discussed in details in Chapters 7 and 8 of this book.

It is believed that knowledge in our brain is represented at the different levels: sub-symbolic, symbolic and non-formal, and symbolic and formal. In artificial intelligence, we can distinguish following structures in which knowledge can be represented: production rules, semantic nets, frames, or scripts. The most important part of the knowledge implementation process is to learn the different kinds of the knowledge from the different domains and to utilize the knowledge during SUS activity. In order to ensure the compatibility of learned knowledge there is a need to search for the suitable knowledge representation. When the most of existing knowledge representation forms can be applied in the knowledge implementation, the real challenge is to find the way how to "blend" visual knowledge with non-visual knowledge and how to ensure the compatibility of the knowledge from the different domains or different areas of our human experience. In our approach (knowledge representation) the visual knowledge is represented in the form of the symbolic names and the visual concepts. The learned non-visual knowledge that is needed to interpet the text is represented by the query form, the basic-form, the procedural-form, the explanatory and the interpretational script. Learning of the visual and nonvisual knowledge as the part of the knowledge implementation will be described in the following Chapters.

According to Piaget, learning knowledge about an object starts in the early childhood (infancy) when acquired child knowledge is based on interactions and mobility that allows child to learn new things. In the period of child development intelligence is present, motor activity are developed but no symbols, limited knowledge is developed and some language skills are developed at the end of this stage. Machine learning methods offer only limited means to cope with learning of the knowledge of visual or sensory objects. Most of these methods are focused on recognition of the small group of objects or objects extracted from images. Before our work [1], there was no method that could cope with problem of representation of the visual knowledge. There was no method that could combine representation of the visual knowledge with representation of the non-visual knowledge. In the knowledge implementation approach learning of the new knowledge that comes from an experience of perceiving of the world is connected with acquiring knowledge about visual objects. Learning knowledge of the visual objects, described in Chapter 8, is focused on the proper naming of the perceived object and acquiring basic knowledge about visual categories. Learning about sensory objects requires to learn the concept of sensory object based on data from simple sensory attributes such as colour or shape but also to apply complex measurement tools to gather data. In parallel to acquiring of new knowledge, SUS needs to develope new skills. Acquiring new perceptual skills of SUS needed in perceiving the world are connected with implementations of the complex image processing algorithms and learning of the complex patterns of the visual resoning sequences. Understanding that depends on acquired knowledge and skills links the perceived object through its name with all previously acquired knowledge. Understanding of the object, which belong to the category of visual objects, requires engaging different skills and knowledge. For example, objects from the different structural categories such as the element category, the pattern category or the picture category, require application of different skills and knowledge for processing, for visual reasoning and for visual understanding. An object, which belongs to the category of visual objects, can be named (recognized) based on its form (shape). Also, members of the category of symbolic signs or visual symbols can be named (recognized) based on shape of the object. In the case of members of the category of symbolic signs or visual symbols, visual understanding plays a key role in naming of these objects. In contrast, naming of objects that are members of the category of text involves a smaller part of visual knowledge but at the same time requires a vast amount of non-visual knowledge during understanding process. Also members of the pattern category, derived from the category of visual symbols such as the category of mathematical expressions, the category of texts, the category of musical texts or the category of engineering schemas, need to be interpreted by using very complex methods requiring huge amount of non-visual knowledge. Understanding of the complex visual object is to know from which part it consists of, what material is it made of, what for is it used for, how to make it, how to use it and how does it work.

As it was shown in Chapter 1, in contrast to both animal and machine learning, human learning occurs as part of education, personal development, school or training and formal learning is learning that takes place in a school system. At each educational level learned knowledge is designed to meet criterion of ability of student to learn. Following the way of learning at a school, SUS learning is controlled by the difficulty factor that means, at first SUS learns to comprehend simple texts and solve simple problems that are suitable for a student from the year 1 to 3. Next SUS is learning knowledge that is suitable for a student from the year 7 or high school level. When, for example, year 2 student will have a big difficulty to learn the advance knowledge from the selected knowledge domain such as mineralogy, SUS can understand text from this area however under condition that it has some formal style.

As it was shown at the beginning of this Chapter, the knowledge base of the existing knowledge-base systems is being incrementally developed via a cycle of testing and debugging and isolating a specific bug may require a trace through dozens of inferences involving hundreds of facts and it is a very time consuming. Similarly, SUS learns incrementally via a cycle of learning a new knowledge and testing new knowledge in the context of all learned knowledge. However, in comparison to the human learning SUS can be developed to obtain very complex ability to understand the different problems and the different aspects of the real world as well as some aspect of the cultural world. While the development of SUS can be very time consuming, learned knowledge can be passed from one copy of SUS to another without any need for further learning. Each new developed version of SUS and all its copies will have the same understanding ability.

As it was described in Chapter 1, learning at a school is based on a curriculum program. The learned subjects are based on the specific features of academic discipline. An academic discipline is a branch of knowledge, which is formally taught, either at the university, or via some other such method. Each discipline usually has several sub-disciplines or branches, and distinguishing lines are often both arbitrary and ambiguous. Examples of broad areas of academic disciplines include the natural sciences, mathematics, computer science, social sciences, humanities and applied sciences. Learning knowledge that is part of the learned subject such as mathematics, physics, chemistry, biology, geography is focused on limited scope of the knowledge of a given domain. Most of the concepts used in these domians are well defined and have very clear meaning. The reald world situation or phenomenon is usualy represented by a mathematical model that is relatively easy to understand. There are subject such as mathematics, physics, chemistry, biology, geography, history or English. The curriculum is based on a more general syllabus, which specifies what topics must be understood, and to what level to achieve a particular grade or standard. Existing machine learning methods usually do not refer to the subjects or topics that are part of the school curriculum program. In contrast to existing machine learning metods, in our approach (the knowledge implementation) SUS learning is based on the curriculum program of the schools. SUS learns knowledge of subjects such as mathematics, physics, chemistry, biology, geography, history or English following general syllabus, which specifies what topics must be understood, and to what level to achieve a particular grade or standard. For each subject such as physics at each educational level the topics prescribed by the syllabus are selected and knowledge for each topic is learned. However, in contrast to human learning SUS can learn, in some cases even advanced knowledge withaut learning of the basic knowledge. For example, SUS can learn mineralogical knowledge that usually students learn at the teritiary educational level.

As it was described in Chapter 1, learning needs to be evaluated in order to find if the learning achive its goal. Evaluation of the learning process is usually focused on the individual learner and in order to monitor the results of learning the different methods of testing are used. Tests offer useful information about students' achievements as well as general ability of students and there are tests for each learned subject such as mathematics, physics, chemistry, biology, geography, or history. Also the existing knowledge-base systems are tested and debugged during incrementally developing process. SUS is tested in the context of learning of new skills as well as in the context of the understanding ability to understand object from the visual category and the ability to solve problems from the selected topics subject such as mathematics or physics. In the context of the knowledge implementation testing ability of SUS to learn new skills and new knowledge will be call SUS assessment. The important part of SUS assessment is measurement of the ability of SUS to solve tests that are used for testing student performance in school and testing a general ability of student.

In order to increase students performance on the test the special means for helping with difficulties in learning are used. If a student finds he continually has difficulty in certain area, then there is a special Basic Skills book that will take the student through the topic step by step. Similarly like student SUS will have difficulty in area where SUS does not learn proper skills and knowledge. In order to improve the result of learning in these areas there is a need to learn to solve many prolems that will cover material that student is learning. In order to "perform well" on the test from selected subject such as physics SUS needs to learn to solve many different problems for each specific topic.

Knowledge is transferred to a student by applying the different learning/teaching methods. Scientific knowledge is knowledge that is the result of scientific research and that is stored in the form of scientific books or scientific journals. Knowledge of subjects learned at a school, such as physics, is based on knowledge of one of the branches of knowledge. Knowledge is produced by scientists and stored in scientific institutions such as laboratories, scientific institutes or libraries. Modern knowledge-based systems learn knowledge that is usually extracted from human experts. However, human experts have often a problem in describing their knowledge in terms that are precise, complete and consistent enough for use in a computer program. Such knowledge is often subconscious and may be approximate, incomplete, and inconsistent. In comparison to the knowledge extracted from human experts, knowledge that is part of the curriculum program approved for learning at school for subjects such as mathematics or physics is rather complete and consistent. Learning knowledge that is part of the learned subject such as mathematics, physics, chemistry, biology or geography, is focused on limited scope of the knowledge. Most of concepts used in these domains are well defined and have ruther clear meaning. The real world situation or phenomenon is usualy represented by mathematical model that is relatively easy to understand. However, subject such as English refers to general knowledge where concepts are fuzzy and does not have unique meaning. The liteature text from subject such as English can refer to very unique world situation or even fictious facts and there is not easy to evaluate to which extent the text is understood. SUS is learning knowledge that is part of the curriculum program approved for learning at school and by this is avoiding problems connected with knowledge extracted from human experts. It is however, possible that during SUS development when SUS will be able acquiring new knowledge at the higher educational level that will be a need to extract knowledge from human experts.

Knowledge-based systems acquire knowledge from sources of the knowledge such as textbooks and other documented knowledge or human experts and their experience in the problem domain. As it was described in Chapter 1, learning at school is based on books such as handbooks or books that help to improve the learned results. Many books offers solved and supplementary problems that are aranged to allow a progression within each topic and some books ofers the answer for the solutions or even can offer a complete step-by-step explanation for the solutions. A complete step-by-step explanation for the solutions is very useful during SUS learning. Learned step-by-step explanation for the solutions make it possible to learn the ability to provide an explanation for solved problems. It should be noted that to understand and to solve problem does not mean to be able to provide an explanation for solved problem. Books used during learning at school usually cover the material that is subject of learning at primary, secondary, or teritiary educational level. The books used at primary level refers to the simple knowledge. If a student finds he continually has difficulty in certain area, then there is often a basic skills book that will take the student through the topic step by step. Those books contain solved problems or additional exercises for students, the aim of which is build and reinforce basic skils in reading, comprehension and mathematics. The solved problems that are often presented in those books are of two kinds: those that show how numerical answers are found to typical questions, and those that review important facts and ideas. A collection of multiple-choice exercises supply a test of the reader understanding ability of the learned content and the supplementary problems give the student a chance of practice what he has been learned. Very often, for greater clarity of presented material, problems are carefully selected, explanations are included and figures are added.

During SUS learning the main source of the knowledge of the selected subject are books that are used by students at school such as handbooks or books with suplementary materials. Important books used during SUS learning are suplementary books with solved problems of two kinds: those that show how numerical answers are found to typical questions, and those that review important facts and ideas and multiple-choice exercises that test understanding of the text contents. The books used for SUS learning can be grouped into one of the following classes: type of books (eg. Handbooks), the type of educational level for which books cover the material that is subject of learning (primary, secondary, or teritiary educational level), year of schooling (e.g. books for years 1 to 7), reference to the student skills (e.g. year 2), type of learned subject (e.g. mathematics, physics), type of topic (e.g. numbers, measurement and shapes in math), aim - to build and reinforce basic skills in (e.g. reading, maths), aim- allowing students to learn new concepts, type of problems (e.g solved and supplementary problems), arangment of problems (e.g. to allow a progression within each topic), type of problem-solving strategy (e.g. one model problem or many problems following the model), problems goal (e.g. to practice the skills in solving a given problem), having or not an answer for solution, having or not a complete step-by-step explanation for the solutions. Based on these classes, the problems that are used during SUS learning can be selected taking into account factors that can have influence on the learning process.

Human learning is slow because of errors disappearing only gradually. Learning is repetitive as a human needs repeated attempts before a solution path is mastered. SUS is learning during incremental learning process and the learning process is time consuming, however each new version of SUS has improved learned knowledge and learned skills that reflects the ability of SUS to be assessed as to achive the higher educational level (e.g. yer 4). Each copy of a given version of SUS has the same ability to understand. Humans continue to improve their prformance over long period of practice. For example, to ensure understanding of the concepts of chemistry, mathematics or physics there is a need to solve many problems on each topic. Similarly the knowledge implementation assumes learning to solve many problems on each topic.

Learning depends on what other problems have been solved. There are numerous methods to solve most problems. The solution methods are usually related to each other but may seem very different. Some related problems can be solved using one method, and some with another. Many of the problems use several solution methods. Learning of the different methods can be useful for solving similar problems. SUS needs to learn to solve many problems on each topic. The important part of the knowledge implementation is to find solution methods that are related to each other. For each problem that has more than one solution method all solution methods are learned. During understanding process the one solution method is selected based on criterion of selection the simplest one. The very important part of the knowledge implementation is to learn to identify the similar problems. Problems are regarded as similar, in the context of the similarity at the query-form level, basic-form level or procedural-form level. The meaning of the terms the query form, the basic-form, or the procedural-form will be explained in the following Chapters.

The proper learning of solving problems is to start with the more basic ones and progress to those that are most difficult. Each problem needs to be read carefully since a small difference in the wording of a problem can make a large difference in its solution. The knowledge implementation (SUS learning) starts with the more basic ones and progresses to those that are most difficult. Important part of the knowledge implementation is to order learned problems from the easy one to the most difficult ones.

In constrast to human learning, SUS can learn a large number of categories or problems without forgetting some of them. For example, names of all known minerals or their characteristics such as chemical compositions can be learned as facts that are easily accessible and once learned they will never be forgotten. In the case of an error during learning process the error can be easy to correct during testing process. For example, learned incorrect name of the mineral can be corrected during testing process by exchanging the incorrect name with the correct one.

Learning by human is always accompanied by learning contextual noise and envinronmental context. Learning is performed in a given day, month, year, in a given age of learner, in special circumstances such as school, in a given room with a given aesthetic inside, with clasmatess. Many of this information are learned as the envinronmental context and this information is not relevant to the contents of learned material. This historical record that can be remembered as some vivid facts does made a very specific learning process that can be only characteristic for human.

Human being is not a machine that can learn only knowledge that is needed for the labour market but has its own specific needs such as the contemplation of the beauty of the world or being angaged in love. Never the machine will be able to acquire our human ability to love and be loved.

4 Understanding and Learning

In the previous Chapters, the knowledge implementation was defined in the context of both human learning and machine learning. The knowledge implementation is focused on learning of the knowledge and skills by machine (SUS) and is concerned with two main aspects of human learning: learning of the visual knowledge in the context of the categorical structure of the learned categories of the visual objects and learning of the knowledge that is connected with understanding of the content of the text. As it was described in Chapter 3, SUS operates in two main modes, learning and understanding mode. SUS ability to understand depends on the effectiveness of learning process and learning of the new knowledge depends on SUS ability to understand.

It is understanding that sets man above the rest of sensible beings, and gives him all the advantage and dominion which he has over them. Understanding and learning are terms that describe the main human intellectual activities. Understanding of acquired knowledge is the basis for meaningful learning. Machine understanding is the term introduced by authors to relate this term to the term machine learning used to describe the process of acquiring knowledge by machine. Machine understanding is part of the knowledge implementation that stresses dependence of the learning and understanding process. SUS as a machine that is designed to have the ability to think and understand, needs to learn both knowledge and skills. In this Chapter learning is regarded, within framework of the knowledge implementation, as the process of supplying of the material for thought that leads to understanding. As it was described in [1], the main ingredient of thought is a concept. The concepts are part of hierarchical structure of conceptual knowledge that is learned and transformed into the form that is accessible during thinking process. In this book the term category is often used to denote the term concept. The term category is used to denote the class of visual objects that are part of the real world, whereas the term concept is used to denote members of the class with reference to thinking/understanding process. Similarly, in area of philosophical investigations, the term category is used to stress the difference between the concepts that have connection with human mind and the category that is an object that is the result of the generalization process. Understanding appears as the result of the thinking process and is based on abilities called intelligence such as a verbal communication, spatial orientation, memorizing, and reasoning. Understanding that is based on knowledge is often connected with interpretation or disclosing meaning of the language and the concept is the key element of understanding process. Understanding and thought were topics of many philosophical thinkers such as Plato, Aristotle, Locke, Berkeley, Laibnitz or Gadamer to list just

Z. Les & M. Les: Shape Understanding System - Knowledge Implementation and Learning, SCI 425, pp. 27–32.
springerlink.com

a few (see. e.g. [10], [98], [99], [100], [101], [102], [103]) and were regarded in the context of the origins of human knowledge.

Visual perception was often thought of as the 'introduction' to understanding of the real world phenomena. For example, Locke claimed that perception is made up of sensations (input) and reflections, Wittgenstein underlined the role of the knowledge and Arnheim and Rock [104] suggested that perception is intelligent in that it is based on operations similar to those that characterize thought. According to the inference theory, knowledge is acquired solely by sensory experience and association ideas and that the mind at birth is a tabula rasa upon which experience records sensations. The Gestalt view was far from being a tabula rasa and refers to the laws of association: items will become connected when they have frequently appeared together [105]. The Gibson theory [15] claims that sensory input is enough to explain our perceptions. Marr [16] developed a primal sketch paradigm for early processing of visual information. SUS understand of the visual object by transforming it into the symbolic name and next during naming process the name is assigned to the perceived object. The processes connected with naming of the object are related in some aspects to human perception. Understanding of the visual object or phenomena is different from understanding a general concept, or abstract concepts such as mathematical objects and the visual concept and mental images play a key role in visual understanding.

Learned knowledge is used during understanding of the visual or sensory object and during understanding and interpretation of the text object. Understanding of the object, which belongs to the category of visual objects, require engaging different skills and knowledge. For example, objects from the different structural categories such as the element category, the pattern category or the picture category, require application of different skills and knowledge for processing, for visual reasoning and for visual understanding. Understanding of the complex visual object is to know which part it consists of, what material is it made of, what for is it used for/by, how to make it, how to use it and how does it work. An object, which belongs to the category of visual objects, can be named (recognized) based on its shape. In the case of members of the category of symbolic signs or visual symbols, visual understanding plays a key role in naming of these objects. In contrast, naming of objects that are members of the category of text involves only a small part of visual knowledge but at the same time requires a vast amount of non-visual knowledge during understanding process. Also members of the pattern category, derived from the category of visual symbols such as the category of mathematical expressions, the category of texts, the category of musical texts or the category of engineering schemas, need to be interpreted by using very complex methods requiring huge amount of non-visual knowledge. The symbolic name and visual concept can explain the way in which the pictorial information can be represented, however in this book these issues will not be discussed.

Understanding of the sensory objects, similarly to the visual objects, requires accurate naming of the object. Understanding both the visual and the sensory objects begins with the naming of the object. When an examined object is named correctly (recognized) all previously acquired knowledge can be used during the understanding process. In the case of naming an object from the mineral-category, measurement of many different minerals' properties, such as a specific gravity or obtaining a complex data such as the data from X-ray analysis, is required. The reason is, that some visual attributes of the object, such as a color or a form (shape), supply not enough information to provide a reliable naming of that object. There are non-visual sensory objects such as speech objects that require a huge amount of non-visual knowledge during understanding process. For practical reasons, in the context of understanding by SUS, a speech object is transformed into the text object and interpreted as the text object.

Understanding of a text involves complex processes connected with interpretation of the different texts and finding their meanings. The text used in educational process are called the educational text. Understanding of the the educational text is related to the different scientific domains such as geography or chemistry. Solving problems given in the form of the mathematical of physical tasks, that are part of the educational text is the ultimate proof of understanding and competence. The educational tests that are used to test ability of students to understand are applied to test understanding abilities of SUS. Understanding requires to have a very well developed problem solving skills. The most important skills in problem solving are skills in solving educational tasks. Solving educational tasks is performed in two main stages: the first is to understand a task and the second one is to solve the task. Understanding of the task is made at two different levels. At the first level the real world situation is reconstructed whereas at the second level the type of the problem is identified and solved.

Understanding is related to the scientific discipline that defines the scope of interest of a given discipline. For example, physics is about understanding how everything works, from nuclear reactors to nerve cells to spaceships. By "understanding" it means that we can predict what will happen, given a certain set of conditions. Usually theories are based on simplified models of real physical system. New theories in physics do not invalidate earlier ideas, but they instead extend their range of applicability. Application of idealized models of nature makes it possible to avoid the overwhelming complexity that makes many physical systems intractable to mathematical analysis. For example, study of the behavior of projectiles ignore an air effect. Although this is not realistic, it is a quite good approximation in many cases, and it gives us good insight and understanding of the physical process. Understanding and competence in physics means to be able to solve problems. Understanding physics means to take into account the specific of this discipline. Physics has a special vocabulary that constitutes a language of its own, a language immediately transcribed into a symbolic form that is analyzed and extended with mathematical logic and precision. Words like energy, momentum, current, flux, inference or capacitance have very special scientific meaning. These must be understood correctly because the discipline builds layer upon layer. In solving problem there is very important to read each problem carefully, since a small difference in the wording of a problem can make a large difference in its solution. In order to solve the problem there is a need to understand many concepts that are not necessary relevant in the context of the physical categories. For example, in the task T0 "student driving car travels 10.0 km in 30.0 min. What was his speed?" the information "student driving car" and "travel" is not relevant for finding the solution, however during solving this task we need to understand why this information is not relevant. In order to solve the task some information needed could be inferred from the content of the task. For example, in order to solve the task T0 assumption was made about the type of movement.

Understanding physics is to "perceive" an object or the phenomenon as a model that can be represented by the schematic graphical form or mathematical formula. During solving physical task we are looking for the mathematical model or the schematic representation of the phenomena that can explain some aspects of real world processes. The physical knowledge consists of definitions that refer to the physical models and are formulated in terms of physical categories. The definition can be given in the form of linguistic description, for example, "Machine is a device by which the magnitude, direction, or method of application of a force is changed so as to achive some adventage". The visual schematic representation is often used in physics in order to understand the principle of "how it works". For example, the lever can be defined as "lever is simple machine used to amplify physical force" and represented as shown in Fig. 4.1. The amplification of the physical force can be computed from the following rules:

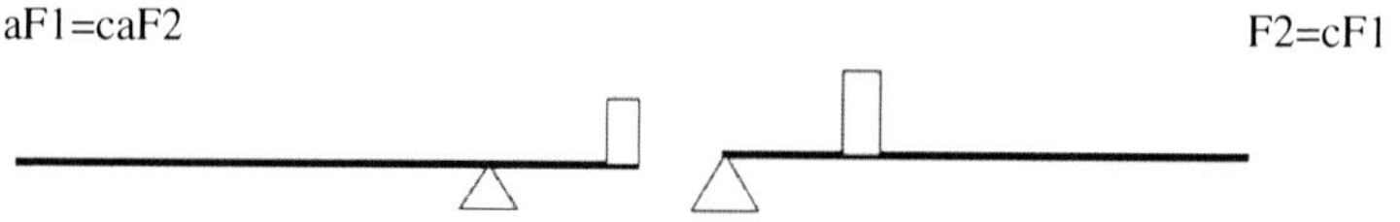

Fig. 4.1. Example of the visual representation of the lever

In the cases when a model is very useful in finding a solution and explaining some aspects of the real world there is a tendency among some scientists to misunderstand the model assumptions and take an abstract model of the object to be an object itself. We would like to stress that all physical categories refer to the differend models of the real world and by this all physical descriptions are concerned with the description of the differents models of the real world.

For chemistry, the best way to ensure understanding of the concepts of general chemistry is to solve many problems on each topic. A large number of different problems, rather than revoking the same problems again and again need to be attempted. In the case of revoking of the same problems the solution can be memorized. The proper learning of solving problems is to start with the more basic ones and progress to those that are most difficult. There are numerous methods to solve most problems. The solution methods are usually related to each other but may seem very different. Some related problems can be solved using one method, and some with another. Many of the problems use several solution methods.

Understanding of the mathematics is concerned with understanding of the mathematical concepts and mastering skills in problem solving in order to solve mathematical tasks. Also a definition, a theorem, a colloraly and a lemma are very important parts of understanding of the mathematics. The mathematical definition is given in the form of linguistic description or in the form of the symbolic expression. The symbolic expression needs to be translated into the linguistic description. For example, the definition given in the form of the symbolic expression $X \subset Y \equiv \prod(x \in X \rightarrow x \in Y)$ can be translated into the following linguistic expression: "the set X is included in the set Y (set X is a part of the set Y (the set X is a subset of the set Y) if and only if every element of the set X is an element of the set Y". The important part of the mathematical problem solving skills is ability to construct mathematical proof. The proof is a sequence of sentences (formulae) which is called proof lines. The mathematical proof can be constructed by applying the axiomatic methods or mthod from assumption. Constructing mathematical proof is to follow the rules and utilize the knowledge given in the form of definitions.

Another important part of understanding of mathematic is to understand of the methods of solving of the mathematical text-tasks. The mathematical text-task is the task that can be transformed into the symbolic mathematical expression. The symbolic mathematical expression describes the rule of finding of the solution. The basic type of task can include explicit type of equation, for example, "solve equation x + 4 = 7" which is transformed into "solve equation L1(x)", whereas "solve equation x^2 +4x+ 4 = 7" is transformed into "solve equation L2(x)". The mathematical task can be given in the form of the description of real world scenes. For example, "Lana bought 2 apples at 35c each, 3 bananas at 55c each and 5 mandarins at 25c each. Which one of the following would calculate the total cost in dollars of what Lana bought?" or "Danny bought a salad sandwich, a fruit juice and ice cream from the canteen. How much change should he get if he paid with 10$".

Computer science has its own specific methodology and categories. Understanding computer science is to be able to find an algorithm how to solve it. The term algorithm is used in computer science to describe a problem-solving method suitable for implementation as a computer program. Most algorithms involve complicated methods of organizing data used in computation. Object created in this way is called data structure. When a large computer program is to be developed a great deal of effort must go into understanding and defining the problem to be solved, managing its complexity and decomposing it into smaller subtasks that can be easy implemented. Implementing simple versions of basic algorithms help us to understand them better and thus use advanced versions more effectively. The recommended strategy for understanding program is to implement and test it.

SUS learns to understand the objects that belong to the different categories. The very important role in the understanding of the visual or sensory object plays the correct naming of the object. In contrast, finding the meaning of the text plays a key role in the understanding of the text objects. Finding the meaning of the text is based on the interpretation of the content of a text.

To evaluate the ability of SUS to understand the text, parts of standardized student achievement tests are applied. It is assumed that SUS understands the text from the selected text category if SUS can solve the different tests such as standardized student achievement tests. Learning of the text requires transferring knowledge from the different available sources and means into SUS. Similarly to a student, SUS learns different texts that belong to the different text categories to perform well on the student achievment test. The student achievement test is a test used for monitoring students' perforamnce at school. This type of tests are used to assess the ability of SUS to understand learned knowledge.

5 Shape Understanding Method

In Chapter 3, the knowledge implementation is defined in the context of both human learning and machine learning. The term the knowledge implementation introduced in this book is refering to learning of the knowledge and skills by the machine, the shape understanding system (SUS). It is concerned with two main aspects of human learning: learning of visual knowledge in the context of the categorical structure of the learned categories of the visual objects and learning of the knowledge that is connected with understanding of the content of the text. The knowledge implementation is part of the shape understanding method, developed by authors [1] that stresses the importance of the knowledge acquisition in understanding and thinking processes. In this Chapter some aspects of the shape understanding method, that allow for better understanding of the knowledge implementation approach, are presented.

When the mind makes a generalization such as the concept of a tree (Fig. 5.1), it extracts similarities from numerous examples. The simplification of the complex form by making generalization enables higher-level thinking. Similarly, the shape understanding system (SUS) acquires knowledge by extracting similarities from the numerous visual objects in order to form a visual concept.

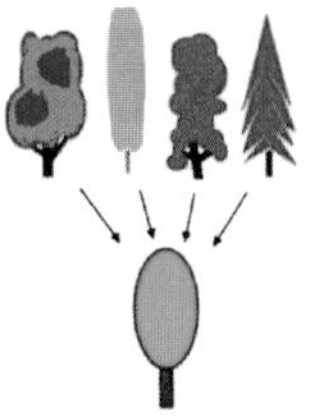

Fig. 5.1. Example of generalization of the concept of a tree

The shape understanding system understands an examined object based on knowledge and skills that were learned previously. Based on the general properties of objects that are part of the real world, the different epistemological categories of objects were established. The category of the visual objects is the basic epistemological category, described in Chapter 6. From the category of the visual objects, the category of the sensory objects, the category of the text objects and the category of aesthetic objects were derived (see Fig. 5.2). The category of the text objects, that is decribed in Chapter 7, is the category of visual objects that have dual meaning. The category of the aesthetic objects is the category of visual objects that are the subject of the aesthetic evaluation based on the aesthetic attributes of these objects.

Z. Les & M. Les: Shape Understanding System - Knowledge Implementation and Learning, SCI 425, pp. 33–43.
springerlink.com

a b c d

Fig. 5.2. The examples of (a) the visual object, (b) the sensory object, (c) the text object and (d) the aesthetic object

In this book the visual objects, the sensory objects and text objects are described in the context of how SUS learns to understand objects that belong to these categories. The aesthetic evaluation of an object will be a topic of further study, based on the framework described in [106], [107].

For the purpose of this study, the text category T is divided into four different specific categories: the text-query category T^G, the text-task category T^T, the dictionary-text category T^D and the long-text category T^L. The text-task category is divided into the category of visual-text-task $T^T[V]$, the action-text-task $T^T[A]$, the explanatory-text-task $T^T[W]$, the educational-text-task $T^T[E]$ and the IQ-text-task $T^T[I]$. In this book only the text-query category T^G, the text-task category T^T and the dictionary-text category T^D is described.

SUS learns to understand the objects that belong to the different categories. The very important role in the understanding of the visual or sensory object plays the correct naming of the object whereas finding the meaning of the text plays a key role in the understanding of the text objects. In order to evaluate the ability of SUS to understand the text-tasks, standardized student achievement tests (selected tasks) were applied. It was assumed that SUS is able to understand the text from the selected text category if SUS can solve the different tests such as standardized student achievement tests. Learning of the text requires transfering of the knowledge from the different available sources and means, such as dictionaries or handbooks into SUS. Similarly to a student that is learning for passing a test, SUS learns different texts, from different achievemnt tests, that belong to the different text categories in order to "perform well" on student achievment test. The student achievement test is a test used for monitoring students' perforamnce at school. This type of tests are used to assess the ability of SUS to understand learned knowledge.

An object from the text category is interpreted based on the category of body of the knowledge $\langle K_{BodK} \rangle$. The category of body of the knowledge (knowledge category) $\langle K_{BodK} \rangle$ is divided into the category of theology $\langle K_{Teol} \rangle$, the category of philosophy $\langle K_{Phil} \rangle$, the category of science $\langle K_{KSc} \rangle$ and the category of common sense knowledge $\langle K_{KSK} \rangle$. The category of science (the knowledge object) $\langle K_{KSc} \rangle$ is divided into the category of physical sciences, the biological sciences, the medicine, the engineering or the social sciences. Science is any system of the knowledge that is concerned with the physical world and other phenomena and that has its own research methodology to pursue new knowledge. In general, science involves pursuit of the knowledge covering general truths or the operations of fundamental laws. For simplicity, the categories derived from the visual objects will be called the visual categories whereas the categories derived from the knowledge objects will be called the knowledge categories.

Finding of the meaning of the text requires learning vast amount of knowledge. Learning is an iterative process and a learned knowledge is tested and corrected at each iterative stage. To improve the ability of SUS to acquire knowledge during learning the coding categories are introduced. The coding categories c_i are related to both the knowledge categories κ_i and the ontological categories of the visual objects v_i , and they are learned during iterative learning process. These categories are established in order to obtain the generalization of the learned texts and to learn the query-form $\Theta(T^T)$. During understanding of the text T each word $w_i \in T$ is coded as one or more than one coding category from the categorexicon $c_i \in C$. The text transformed into a set of coding categories $c_i^1,...,c_i^N$ is used to find the basic-form, the procedural-form, and the explanatory and interpretational script. The coding categories will be described in more detail in Chapters 7 and 8.

When an object from the text category is understood by finding meaning of the text, an object from the category of visual objects or sensory objects is understood by naming of an examined object. The visual object after naming is interpreted based on knowledge of the ontological visual categories and knowledge of the knowledge schema. Categories of the visual objects are established based on the assumption that a visual object exists and can be perceived by the accessible technical tools [1]. Categories of visual objects supply knowledge about the visual aspects of the world. The notation of basic knowledge of categories is based on a categorical chain. The categorical chain is a series of categories derived from the categories of visual objects or the categories of a body of the knowledge showing the hierarchical dependence of knowledge. The categorical chain derived from the categories of the visual objects is given as follows: $v_O \triangleright \langle \pi \rangle \triangleright \langle \sigma \rangle \triangleright \langle v \rangle \triangleright \langle v \rangle .... \triangleright \langle v \rangle \triangleright \langle v \ , v \ ,....., v \ , \rangle$, where the categories of the visual objects are derived from the category of the visual object v_o. A category at the first level of the categorical chain is called the perceptual category $\langle \pi \rangle$ of a visual object. A category at the second level of the categorical chain is called the structural category $\langle \sigma \rangle$ of a visual object. The ontological category $\langle v \rangle$ begins from the third level of the categorical chain. The symbol $\triangleright$ denotes moving to the next level of the categorical chain. Notation $\langle v \ , v \ ,....., v \ \rangle$ symbolizes different categories at the same level of the categorical chain. Notation $\langle v \ , v \ , v \ ,... \rangle$ means that only selected categories are listed.

The perceptual categories and structural categories are associated with the visual appearance of an object and are represented by visual knowledge. The structural element category can represent both the visual and sensory object. The sensory object is usually the visual object that is named based on the complex sensory data rather than on visual features of the object. The structural pattern category can represent the text or the mathematical expression. The structural elements category, such as a category of figures, a category of signs, a category of letters or a category of real world objects, denoted as $v_O \rhd \langle \Pi \rangle \rhd \langle \Lambda \rangle \rhd \langle v_{\mathrm{Re}O}, v_{\mathrm{Im}O}, v_{Sig}, v_{Let}, v_{Fig} \rangle$, refers to both visual and non-visual knowledge. Knowledge of the specific category derived from the visual category such as the symbol category is learned by SUS at the prototype level. Ontological visual categories have hierarchical structure and at the bottom of each categorical chain is a prototype. The prototype is defined during learning process at the level for which the training exemplars are available. The prototype is represented by all visual representatives of a specific category and it is assumed that learned visual knowledge is covering a visual domain prototype. The visual domain prototype refers to visual knowledge that makes it possible to recognize all visual representatives of the prototype. For example, the visual domain prototype for the category of the capital letter "T" consists of all fonts and handwritten characters of the category of the capital letter "T" (see Fig. 5.3), whereas the visual domain prototype for the category of the font aerial capital letter "T" consists only of the aerial font of the capital letter "T".

Fig. 5.3. Examples of objects from the domain prototype for the category of the capital letter "T"

Understanding of the visual object from one of the ontological categories requires learning of the visual concepts of this category. The ontological category v_i is given by its name n^i and is represented by a set of visual objects called the visual representatives of the category $v_i(o) = \{o_1, o_2, ..., o_n\}$. Visual knowledge of the category v_i is learned as a visual concept represented as a set of symbolic names $\varphi_c = \{\eta_1, \eta_2, ..., \eta_n\}$. It is assumed that a set $v_i(o)$ represents all visual aspects of the category v_i.

SUS performs in two modes, learning and understanding mode. During learning of the knowledge of the visual objects, at first, the representative sample of objects from the category v_i is selected, then for each object, the symbolic name η_i is obtained and finally a visual concept of this category as a set of symbolic names $\varphi_c^j(v_i) = \{\eta_1, \eta_2, ..., \eta_n\}$ is learned. Visual knowledge of the learned visual categories is represented as a set of the visual concepts $\{\varphi^j(\nu_1),...,\varphi^j(\nu_K)\}$, where K is a number of learned categories. During understanding of an object u, the perceived object u is transformed into the symbolic name η and next a learned set of symbolic names is searched to find the symbolic name of the category that was learned previously. This understanding process can be represented as: for i=0 to K if $\eta \in \varphi^i$ then $n^i \triangleright N^i$, where $n^i \in N^i$ is the name of the i-th category and a set $N = \{n^1,...,n^M\}$ is a set of all names of categories to which the object can belong, and M is a number of names in the set N. In the case when M=1, the reasoning is stopped and the name n^1 of the category ν_i is used as the name of an examined object. After naming, all non-visual knowledge, that was previously learned

for the category ν_i, is now accessible and can be used in the thinking/understanding process. In the case when M=0, there is a need to use the generalization process to find the most similar object. In the case when M>0 there is a need to use learned non-visual knowledge to properly classify an object to one of the object categories.

As it was described at the beginning of this Chapter, understanding of the visual object is based on the shape understanding method described in [1]. In this method, the visual object o_i that is perceived by SUS, is transformed by a photographic transformation Δ: $\Delta(o) \Rightarrow u$ into the phantom u that is the 2D representation (e.g. photograph) of the object o_i. The phantom u is transformed into a set of critical points $\amalg$ by the sensory transformation $\Im$: $\Im(u) \Rightarrow \amalg$ and next into a symbolic description in the form of a string $R\langle\amalg\rangle \Rightarrow \kappa$ and finally into a symbolic name $K(\kappa) \Rightarrow \eta$. A set of critical points $\amalg$ is stored as the bitmap in the file F. Fig. 5.4 illustrates the process of perception of the real world object. In the process of perception, the real world object (cylinder) is transformed into the phantom (circle). The phantom is transformed into the digital object o represented by a set of critical points $\amalg$, and next into the symbolic name η_i.

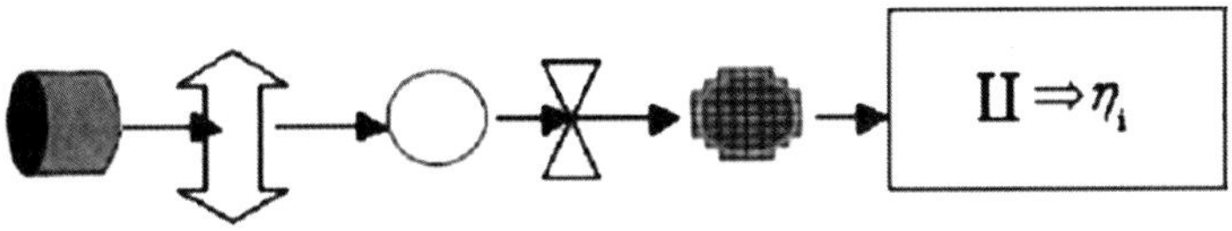

Fig. 5.4. In process of perception, a real world object (cylinder) is transformed into a phantom (circle). The phantom is transformed into a digital object o represented by a set of critical points $\amalg$, and next into the symbolic name η_i.

The symbolic name η_k is extracted from a symbolic description κ_k. The symbolic description κ_k is an intermediate form that has many additional specific data about the perceived phantom. The symbolic description κ_k is used to reason about the specific categories to which the object can belong. For example, the object O1  stored in the file S1.bmp as a bitmap (1024x1024) is transformed into a symbolic description in the form of the following string:

"[A3][[|L3|AE|]|S79||B100,99,99||A60,61,60||G248||@2691|]]{[|L3|O|]|S52||B58,100,57||A29,30,120||G76||@395|]}{[|L3|O|]|S52||B57,100,58||A30,29,120||G76||@396|]}{[|L3|O|]|S53||B100,58,57||A29,120,30||G76||@417|]}".

Next, the symbolic description is transformed into a symbolic name given as a string A3_L3_AE_L3_O_L3_O_L3_O. The symbolic name is used in provisional searching during the understanding process. The perceived object is assigned to one of the shape classes that are represented by a symbolic name. The symbolic name is the name of a shape class and for the perceived object O1 is given as $A\left[L_E^3\right](3L_O^3)$. This symbolic name (shape class notation) is easily transformed into the symbolic name given in the SUS notation as A3_L3_AE_L3_O_L3_O_L3_O.

During the understanding process the perceived object has to be fitted into one of the shape categories (shape classes). A member of the shape class is called an archetype. The archetype ω of the class Ω $(\omega_k \in \Omega)$ is an ideal realization of shape (a visual object) in the two-dimensional Euclidean space (E^2). An exemplar $e \in E$ of the class Ω is a binary realization of an archetype in the discrete space. The exemplar is one of the regions of a binary image. The binary image is regarded as a set of pixels on the discrete grid (i,j). The exemplar $e \in E$ is represented by a set of critical points $\amalg^F = \{p_1^F, p_2^F, \ldots, p_J^F\}$.

During understanding process, a perceived object called a phantom u is transformed into a digital representation given by a set of critical points $\amalg$. The terms 'an exemplar' and 'a set of critical points' have nearly the same meaning $e \equiv \amalg$. The term 'an exemplar' is used to denote that an object, given as a set of pixels seen on the screen, is generated from one of the shape classes in the process called the exemplar generation. The term 'a set of critical points' is used to underline that an object, given as a set of pixels, is the result of transformation during processing stages from one into another set of critical points.

During reasoning process, a perceived object is first transformed into a set of critical points $\amalg$ and next into the symbolic name η. Perceiving an object can be seen as a process of acquiring a new data. In order to fulfill the required task of acquiring the data and processing it in order to obtain a set of descriptors $\Im$, processing methods Φ are used. The processing methods apply an image transformation Θ in order to transform the data into one of the data types. The image transformation Θ is mapping from the one set called the domain of mapping into another called the set of mapping values. As a result of applying the image transformation into a set of critical points $\coprod$, a new set of critical points $\coprod'$, a set of transform numbers Δ or a set of mapping numbers $\sum$, is obtained. The descriptor transformation $\aleph$ is applied to find a set of descriptors $\iota \in \Im$ used to assign the perceived object to one of the possible classes Ω^{η}. A reasoning process that is part of a visual reasoning process is performed passing the consecutive stages of reasoning. During each stage of reasoning a sequence of image transformations is applied in order to find a set of descriptors. The sequence of image transformations $\Theta^{\lambda}_{\coprod} : \coprod \rightarrow \coprod$ that are used in reasoning process can be written as: $\text{ƛ}_{\alpha_1} : \coprod^{\alpha_0} \rightarrow \coprod^{\alpha_1}$, $\text{ƛ}_{\alpha_2} : \coprod^{\alpha_1} \rightarrow \coprod^{\alpha_2}$,…., $\text{ƛ}_{\alpha_M} : \coprod^{\alpha_{M-1}} \rightarrow \coprod^{\alpha_M}$ or as a composite given as $\Theta_{\alpha_1} \bullet \Theta_{\alpha_2} \bullet \ldots \bullet \Theta_{\alpha_M} : \coprod^{\alpha_0} \rightarrow \coprod^{\alpha_M}$, where Θ_{α_M} denotes one of the image transformations and $\bullet$ denotes the sequential operator. The reasoning involves processing by applying one of the image transformations, computation of the descriptors using a descriptor transformation and assigning an object to one of the possible classes.

Acquired visual knowledge is used during the first stage of the understanding process. At the first stage of the visual interpretation, an object is assigned to one of the perceptual categories $v_O \triangleright \langle \pi_{Si}, \pi_{Ld}, \pi_{Co}, \pi_{Sh} \rangle$, next it is assigned to one of the structural categories $v_O \triangleright \langle \Pi \rangle \triangleright \langle \sigma_{El}, \sigma_{Pt}, \sigma_{Pi}, \sigma_{An} \rangle$ and, at the end, it is interpreted as a member of one of the ontological categories $v_O \triangleright \langle \Pi \rangle \triangleright \langle \Lambda \rangle \triangleright \langle v_{ReO}, v_{ImO}, v_{Sig}, v_{Let}, v_{Fig} \rangle$. Ontological categories are part of the categorical structures of the knowledge about the world comprising of the visual object categories and knowledge categories. A visual concept φ_C and a symbolic description κ_k are stored as visual knowledge of the intermediate level of the visual reasoning process. Knowledge of the intermediate level includes a structural archetype A_k, knowledge of the parts P_k, knowledge of the generalization process G_k and knowledge of the linguistic description of the intermediate level L_k. Non-visual knowledge includes the categorical chain H_k, the knowledge schema S_k and interpretational knowledge. The categorical chain H_k, represents knowledge about the world objects.

Understanding of an object is performed at two levels, the intermediate level and ontological level. At the intermediate level of understanding the object is described in terms of the shape classes. The description of the object at intermediate level refers to the symbolic name. For example, for the object , the symbolic name (in SUS notation) "A3_L3_AE_L3_O_L3_O_L3_O" consists of two parts. The first part "A3" gives a general description of the class that means that the object is the acyclic object with three holes. The second part "L3_AE_L3_O_L3_O_L3_O" gives a specific description of the object. The final description of the object, at the intermediate level of understanding, is given in the form of the linguistic description: "acyclic object with three holes". At this level, an object is also described in terms of the structural archetype. At the ontological level, an object is assigned to one of the ontological categories. An assignation to the ontological category is called naming. Naming not only attaches the name to the perceived object but also "connects" the object with all knowledge that is relevant to the name of the object. Many names from different categories can be attached to the same object and naming can be given at many different ontological levels. In order to assign an object to the specific ontological category, information included in a symbolic description is used to obtain the additional data needed in the reasoning process. For example, an object  can be interpreted as a symbol "eye of dragon" when additional relation is established "all three wholes are equal". In the case of the object O1  the size of holes is given in the string form as |S52|, |S52|, |S53|, as the part of the symbolic description. The object O1 can be also interpreted as a mathematical object (solid pyramid) or as a real world object (a model of a pyramid).

Knowledge of each ontological category is part of the conceptual structure of the knowledge about the world. One of the ontological categories used during categorical learning, presented in this book is the sign category. The sign category refers to a visual object meaning of which is based on the system of conventional rules (the code). The category of signs is derived from the category of the visual objects that is given as follows $O \triangleright \left\langle \nu_{\mathrm{Re}O}, \nu_{\mathrm{Im}O}, \nu_{Sig}, \nu_{Let}, \nu_{Fig} \right\rangle$. From the category of signs, the category of 2D signs and 3D signs is derived. The category of 2D signs is divided into the category of visual symbols ν_{VSym} and the category of symbolic signs ν_{VSymS} and is given as: $\langle \Pi \rangle \triangleright \left\langle \sigma_{El} \right\rangle \triangleright\triangleright \left\langle \nu_{Sig} \right\rangle \triangleright \left\langle \nu_{2DSig} \right\rangle \triangleright \left\langle \nu_{SymS}, \nu_{VSym} \right\rangle$. A category of visual symbols is a category of the well defined objects that are used to compose the complex objects (patterns). Basic knowledge needed to interpret a perceived object is given by the knowledge schema. The knowledge schema is learned as part of the knowledge of the learned category. For example, the knowledge schema for the category of convex close polygons includes the visual concept ∂_{ViC}, the name ∂_{Nam}, the definition ∂_{Def} and the method of exemplar generation ∂_{MGe}, and is denoted as: $\left\langle \kappa_{Pol} \right\rangle \triangleright \left\langle \kappa_{NNaP} \right\rangle \triangleright \left\langle \kappa_{ClCoP} \right\rangle \prec \left\{ \partial_{ViC}, \partial_{Nam}, \partial_{MIn}, \partial_{Def}, \partial_{MGe} \right\}$. The name ∂_{Nam} is given in the form of a linguistic expression of the existing languages and is denoted as follows: $\left\langle \kappa_{Pol} \right\rangle \triangleright \left\langle \kappa_{NNaP} \right\rangle \triangleright \left\langle \kappa_{ClCoP} \right\rangle \prec \left\{ \partial_{ViC} \right\} \left\| \nu_{NLan} \right\| \left\langle \nu_{C1}, \nu_{C2}, \ldots, \nu_{CN} \right\rangle$, where

$\langle v_{C1}, v_{C2}, ..., v_{CN} \rangle$ represents the name categories that depend on the selected language category $\|v_{NLan}\|$.

A sensory object is an object from the category of visual objects that is named based on a set of measurements that refer to attributes of the category to which the object is assigned. Naming an object from the category of sensory objects is to classify an object to one of categories of sensory objects. A sensory object that belongs to the category of minerals is assigned to the mineral category based on the measurement of the characteristic minerals features. The category of minerals, described in this book, is derived from the category of non-man made objects given by the following categorical chain: $\langle \Pi \rangle \triangleright \langle \sigma_{El} \rangle \triangleright \langle v_{ReO} \rangle \triangleright \langle v_{Ear} \rangle \triangleright \langle v_{NLiv} \rangle \triangleright \langle v_{NMan} \rangle \triangleright \langle v_{Min} \rangle$. From the category $... \triangleright \langle v_{Min} \rangle$ the specific mineral category is derived $... \triangleright \langle v_{Min} \rangle \triangleright \langle v_M^i \rangle$. The aim of naming (recognition or classification) is to assign the examined object o_i to one of the mineral categories v_M^i based on a set of measurements $m(a^j)$ and finding the mineral category v_M^i for which the measures of attributes of the object $m(a^j)$ are matched with the values of attributes of the mineral category $u(a_i^j)$.

Understanding of the object, a member of the text category, requires only a small part of the visual knowledge but involves a large part of non-visual knowledge in understanding process. The text category is derived from the pattern category. The objects from the text category consist of objects from the element category such as objects from the letter category. The first step in understanding text is to assign an examined object to the text category and find the language category to which an examined text belongs. In most cases this information about an object is given as a contextual knowledge. For example, an object named as a book contains text as part of a book contents. In the case when an examined object is assigned to the text category and the language category, the object is transformed into streams of words using the optical character recognition (OCR) method. The optical character recognition method is the method of automatic recognition of the raster images as being letters, digits, or other known symbols [108]. The adopted OCR method transforms each member of element category into a member of letter category and next into the word category.

The text is interpreted in terms of its meaning. The meaning for the dictionary-text T^D is usually given by the interpretational script $\beta(T^D) \triangleright S$, whereas the meaning for the text-query and text-task is given by both the explanatory script and the basic-form, denoted respectively as $\beta(T^D) \triangleright (J,B)$ and $\beta(T^T) \triangleright (J,B)$. There is a difference between the text-query and the text-task. The text-query T^G can be represented by one word or more than one word and does not have the query-part. The text-task has both the task part and the query-part and is denoted as $T^T = T_U^T T_Q^T$. The query-part T_Q^T is the sentence or part of the sentence that has the query term such as "what" or "who". The text-task has always the query-part.

The text-task T^T can have different forms, can consist of the different categories and can refer to the different phenomena. Meaning of the text-task that refers to the solution of the text-task and is given by the basic-form, is called the basic meaning $\beta^B(T^T) \triangleright B$, whereas meaning that refers to the interpretation of the text in terms of the real world categories, and is given by the explanatory script J, is called the interpretational meaning $\beta^J(T^T) \triangleright J$. Meaning of the text-task consists of the two different parts: meaning of the text that refers to real world situation (phenomena) and meaning in terms of the task that needs to be solved. The first part, the interpretational meaning $\beta^J(T^T) \triangleright J$ is given by the description of the stereotypical situation in the form of the explanatory script J given at the different levels of description that reveals the different levels of the details. The second one, the basic meaning $\beta^B(T^T) \triangleright B$ requires to transform the text into the basic-form B and to identify the type of the solution by transforming it into the procedural-form. A text-task that has the same basic meaning can have the different query-parts having the same task parts $\beta^B(T_{U1}^T T_Q^T) \equiv \beta^B(T_{U2}^T T_Q^T) \triangleright B$, or can have the different task parts having the same query-part $\beta^B(T_U^T T_{Q1}^T) \equiv \beta^B(T_U^T T_{Q2}^T) \triangleright B$ or both $\beta^B(T_{U1}^T T_{Q1}^T) \equiv \beta^B(T_{U2}^T T_{Q2}^T) \triangleright B$. The two text-tasks have the same basic meaning if both refer to the same basic-form $\beta^B(T_{U1}^T T_{Q1}^T) \equiv \beta^B(T_{U2}^T T_{Q2}^T) \triangleright B$. Similarly, the two text-tasks that differ in the query-part $\beta^B(T_U^T T_{Q1}^T) \equiv \beta^B(T_U^T T_{Q2}^T) \triangleright B$ or text part $\beta^B(T_{U1}^T T_Q^T) \equiv \beta^B(T_{U2}^T T_Q^T) \triangleright B$ and refer to the same basic-form B, can have the same basic meaning $\beta^B(T^T)$. The interpretational meaning of the text-task is given by the explanatory script $\beta^J(T^T) \triangleright J$. The interpretational meaning makes it possible to explain why some parts of the text-task are not relevant in solving of the text-tasks. Depending on the knowledge domain to which text-task belongs, the text-tasks can be divided into the mathematical-text-tasks T^{MT}, the physical text-task T^{PhT}, the chemical text-task T^{ChT}, the biological text-task T^{BT} or the geographical text-task T^{GT}. Similarly to understanding of the text-tasks T^T, understanding of the dictionary-text T^D also requires that SUS understands the text contents, is able to explain the meaning of the text and to answer questions that refer to the meaning of the text. Understanding of the dictionary-text is based on the previously learned interpretational script.

As it was described in Chapter 3, the knowledge implementation is concerned with learning of the visual knowledge and learning of the knowledge that is connected with understanding of the content of the text. Learned knowledge is tested during understanding process. SUS operates in the two main modes, learning and understanding. Process of learning consists of acquiring of the new knowledge and learning of the new skills. Learning of the new skills is to learn both the new methods of solving the problem and implementing of the new methods of processing and storing of the acquired knowledge. Acquiring of the new knowledge involves storing the new knowledge in the form of selected knowledge representation. Learning knowledge of the visual object is to learn the visual concept, the categorical chain and the knowledge scheme. Learning knowledge of the sensory object is to learn the sensory concept (model), the categorical chain and the sensory knowledge scheme. Learning knowledge of the text object is to learn the coding categories, the query-form, the basic-form, the procedural-form, the explanatory script and interpretational script.

Understanding of the visual object begins with the naming of the perceived object. Similarly, understanding of the sensory object begins with the naming of an examined object. In contrast, understanding of the text object consists of two parts, naming of the visual object such as a book or a paper and understanding the content of the text. Naming of the visual object that is a member of the text object category, such as a book provide the contextual information during understanding of the text object. Understanding the computational-text-task is to compute result that is usually a number. Understanding the dictionary-scientific-text is to compare the text with remembered facts. Understanding depends on the skills employed and knowledge used. The simplest understanding of the object from the category of the visual object is based on application of categories of the visual objects represented by the categorical chain. More complex forms of understanding utilize both categories of the categorical chain and definitions of the knowledge schema. The complex understanding is based on utilization of the knowledge in the form of models and scripts.

6 Category of the Visual Objects

As it was described in Chapter 1, the knowledge implementation is concerned with learning of the visual knowledge in the context of the categorical structure of the learned categories of the visual objects and learning of the knowledge that is associated with understanding of the content of the text. In this Chapter the categories of the visual objects will be briefly described. This description is based on the material presented in the first book concerning machine understanding [1]. As it was described in Chapter 1, a concept is a key element of the knowledge that is stored in our brain. The knowledge implementation is concerned with learning of the visual knowledge that is stored as a categorical structure of the learned categories of the visual objects. Categories of the visual objects supply knowledge about the visual aspects of the world. A category at the first level of the categorical chain is called the perceptual category $\langle \pi \rangle$ of a visual object, a category at the second level of the categorical chain is called the structural category $\langle \sigma \rangle$ of a visual object. The ontological categories $\langle v \rangle$ begin from the third level of the categorical chain. The perceptual categories and structural categories are connected with the visual appearance of an object and are represented by visual knowledge. The structural element category can represent both the visual and sensory object.

Perceptual categories

The first level of the hierarchy of the categorical chain is called the perceptual category $\langle \pi \rangle$ of the visual object. The perceptual category refers to the visual representation that is the way in which the 3D object is presented as the 2D representative of the 3D object. The perceptual category consists of the silhouette category π_{Si}, the line-drawing category π_{Ld}, the colour-object category π_{Co}, and the shaded-object category π_{Sh} and are represented as $v_o \triangleright \langle \pi_{Si}, \pi_{Ld}, \pi_{Co}, \pi_{Sh} \rangle$. Fig. 6.1 shows examples of the perceptual categories: the silhouette category (Fig. 6.1 a), the line-drawing category (Fig. 6.1 b), the colour-object category (Fig. 6.1 c) and the shaded-object category (Fig. 6.1 d).

a b c d

Fig. 6.1. Examples of members of the different perceptual categories

Z. Les & M. Les: Shape Understanding System - Knowledge Implementation and Learning, SCI 425, pp. 45–59.
springerlink.com

Structural Categories

The structural category refers to the complexity of the visual representation of the visual object. The visual object can be an isolated visual object called an element category, an object composed from the simple elements called the pattern category or a complex visual object composed from the regions that are interpreted as the elements of the different ontological categories $\langle v \rangle$ called the picture category. The structural categories of the visual objects are divided into the element category σ_{El}, the pattern category σ_{Pt}, the picture category σ_{Pi} and the animation category σ_{An}, and are given as the second level of chain categories: $v_O \triangleright \langle \pi \rangle \triangleright \langle \sigma_{El}, \sigma_{Pt}, \sigma_{Pi}, \sigma_{An} \rangle$.

The element category σ_{El} is the category that represents the isolated visual objects of the one of the ontological categories (Fig. 6.2).

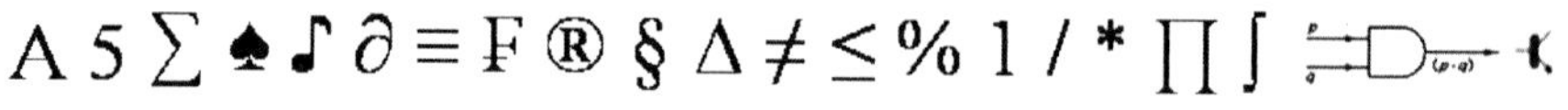

Fig. 6.2. Examples of members of the element category

The category of visual symbols such as engineering symbols or mathematical symbols, that is derived from the structural element category, is given by the following categorical chain: $\langle \Pi \rangle \triangleright \langle \sigma_{El} \rangle \triangleright \langle v_{Sg} \rangle \triangleright \langle v_{VSym} \rangle \triangleright \langle v_{Mth}, v_{Mus}, v_{EnSym}, .. \rangle$, where $\langle v_{Sg} \rangle$ is the signs category, $\langle v_{VSym} \rangle$ is the visual symbols category, v_{Mth} is the mathematical symbols category, v_{Mus} is the musical symbols category, and v_{EnSym} is the engineering symbols category. The category of the electronic symbols such as a resistor, a transistor, a capacitor (see Fig. 6.3), is derived from the category of the electronic engineering symbols.

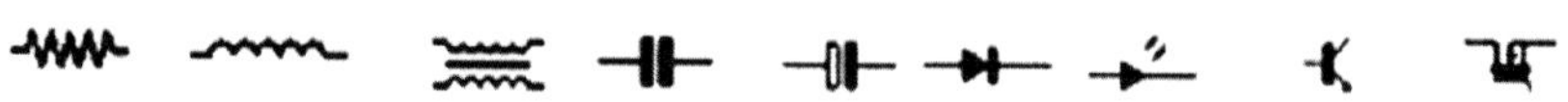

Fig. 6.3. The category of electronic symbols: a resistor, a capacitor, a capacitor electrolytic, a bipolar transistor, a field-effect transistor

Fig. 6.4 shows specific categories of the electronic symbols: a capacitor, a capacitor electrolytic, a bipolar transistor and a field-effect transistor. The category of the engineering visual symbols refers to the real world objects. The electronic elements categories are derived from the category of man-made objects $\langle v_{MMad} \rangle$. The category of the electronic symbols such as a resistor, a transistor, a capacitor shown in Fig. 6.3 is interpreted in the terms of the electronic elements categories shown in Fig. 6.4.

Fig. 6.4. Examples of real world categories: a resistor, a capacitor, a transistor.

The pattern category refers to objects that are composed from the simple components of the element category. Examples of the objects from the pattern category are shown in Figs. 6.5 and 6.6. For example, the pattern category of the visual symbols that is derived from the structural pattern category is shown in Fig. 6.5. From the visual symbol category the following categories are derived: the category of mathematical patterns, the category of musical patterns, the category of coordinate system patterns or the category of pattern of the engineering symbols. Fig. 6.5 shows examples of the pattern categories: the category of mathematical expressions, the category of musical scores and the category of the engineering schemas.

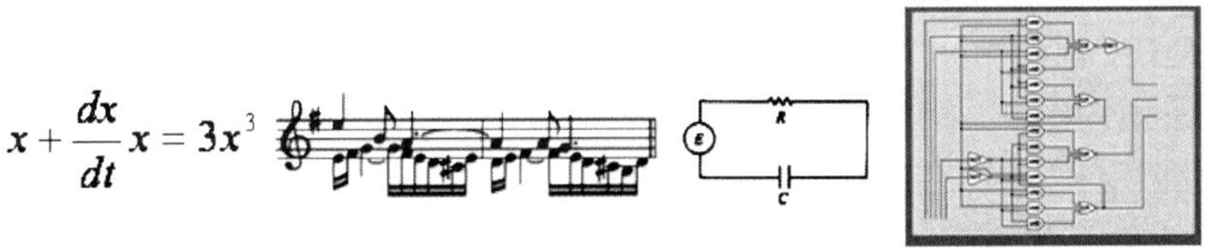

Fig. 6.5. Examples of objects from the pattern category derived from the category of the visual symbols

The pattern category includes such complex visual objects as the engineering drawings, the mathematical objects, maps or the visual intelligence tests (see Fig. 6.6).

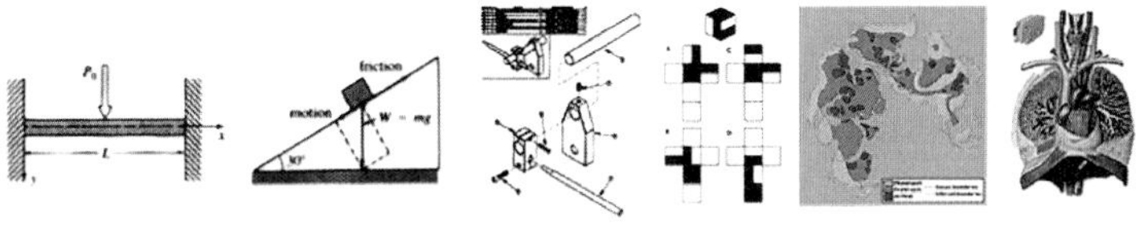

Fig. 6.6. Examples of patterns consisting of elements of the different categories

The category of pictures is derived from the perceptual category of shaded-objects π_{Sh} and is divided into the micro-world category, the macro-world category and the world category and given by the categorical chain as: $v_O \triangleright \langle \pi_{Sh} \rangle \triangleright \langle \sigma_{Pi} \rangle \triangleright \langle v_{\mathrm{Re}O} \rangle \triangleright \langle v_{Mic}, v_{Mac}, v_{Ear} \rangle$. The category of pictures can be divided into categories specified by the perceptual and ontological level given by the categorical chain in the following form $\langle \Pi \rangle \triangleright \langle \sigma_{Pi} \rangle .. \triangleright \langle v_{Mic}, v_{Mac}, v_{Ear} \rangle$. Fig. 6.7. shows a.

the picture of the real world category, b. the picture the micro-world category, c. the picture of the macro-world category d. the picture of the scientific visualization e. the picture of the category of mythological objects.

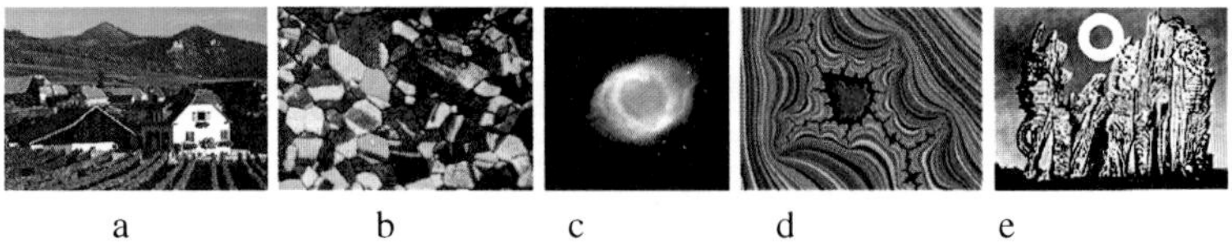

Fig. 6.7. Examples of the different categories of pictures: a. the picture of the real world category, b. the picture the micro-world category, c. the picture of the macro-world category d. the picture of the scientific visualization e. the picture of the category of mythological objects

There is a class of pictures that is distinguished based on their aesthetic quality. Fig. 6.8 shows examples of works of art that are members of the landscape category.

Fig. 6.8. Examples of the works of art that are members of the landscape category

The category of animation refers to the category of processes. Fig. 6.9 shows examples of the objects from the animation category.

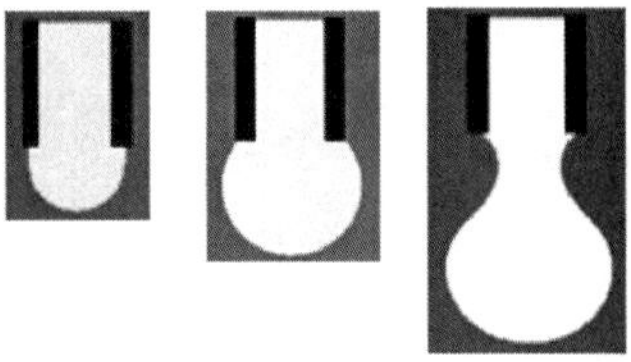

Fig. 6.9. Example of members of the animation category

Ontological categories

The third categorical level is called the ontological level and refers to the meaning of a visual object. The ontological level includes ontological categories $\langle v \rangle$ such as the category of real world objects $\langle v_{\mathrm{Re}O} \rangle$, the category of imagery objects $\langle v_{\mathrm{Im}O} \rangle$, the category of letters $\langle V_{Let} \rangle$, the category of signs $\langle V_{Sig} \rangle$ and the category of figures $\langle V_{Fig} \rangle$. The category of figures consists of the category of 2D figures $\langle V_{2DF} \rangle$, the category of 3D figures $\langle V_{3DF} \rangle$ and the category n-D figures $\langle V_{M\,3DF} \rangle$ (more than 3D figures). The category of 3D figures refers to geometrical objects that 'exist' in the three dimensional space. The category of n-D figures is the category members of which are objects that can be found in more than three-dimensional space. (see Fig. 6.10).

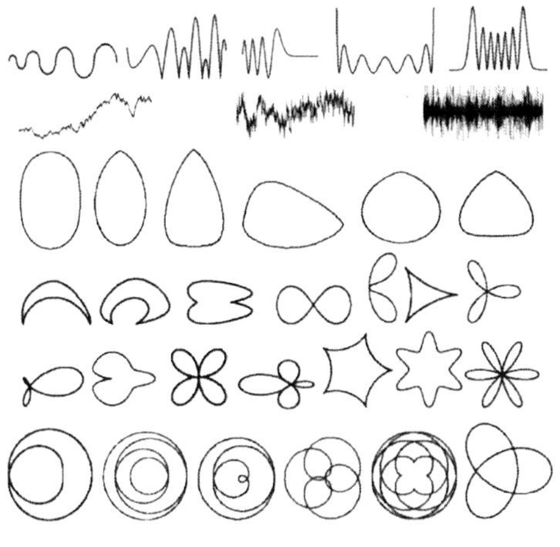

Fig. 6.10. Members of the curve category

The category of 3D figures consists of the category of 3D curves, the category of 3D surfaces and the category of 3D solids. From the category of 3D solids the category of polyhedrons is derived. Fig. 6.11 shows examples of the named polyhedrons.

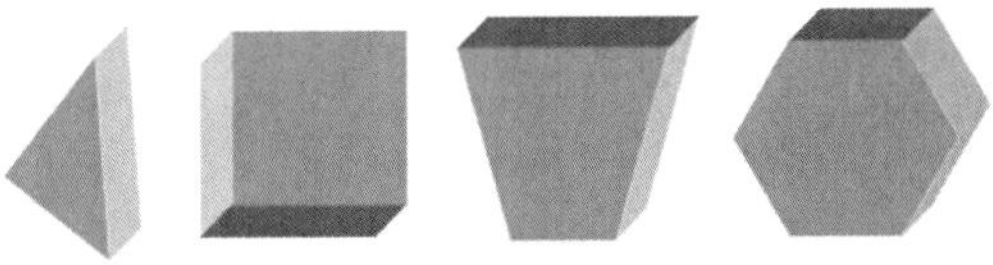

Fig. 6.11. Examples of members of the polyhedrons category

In literature terms a sign, a letter and a symbol are not well defined and are often used as synonyms. In contrast to the letter, the sign is not part of the system of any existing language. Examples of signs are shown in Fig. 6.12.

Fig. 6.12. Examples of signs

The category of visual symbols is the category of the well defined objects that are used to compose the complex objects (patterns). Examples of the category of visual symbols v_{VSym} are shown in Fig. 6.13.

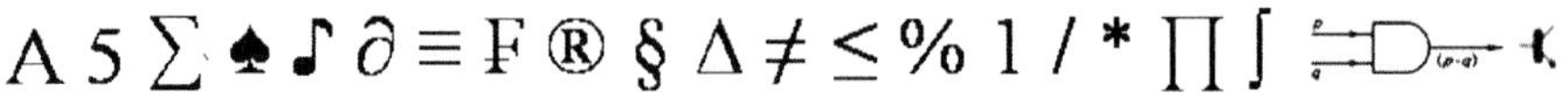

Fig. 6.13. Examples of members of the structural element category and the ontological category of visual symbols

From the category of visual symbols the category of mathematical symbols v_{Mth}, the category of logical symbols v_{Log}, the category of musical symbols v_{Mus}, the category of currency-symbols v_{Cur}, and the category of engineering symbols v_{EnSym} are derived and are given by the following categorical chain: $\langle \Pi \rangle \triangleright \langle \sigma_{El} \rangle \triangleright \langle v_{Sg} \rangle \triangleright \langle v_{VSym} \rangle \triangleright \langle v_{Mth}, v_{Log}, v_{Cur}, v_{Mus}, v_{EnSym}, .. \rangle$.

In contrast to the members of the category of visual symbols, members of the category of symbolic signs cannot be used to compose any complex meaningful object. The meaning of elements of the category of symbolic signs does not depend on the meaning of other elements of the same category. From the category of symbolic signs the category of trademark signs v_{TrS}, the category of road signs v_{RoS}, and the category of cross signs v_{CroS} is derived and is given by the following categorical chain: $O \triangleright \langle v_{Sig} \rangle \triangleright \langle v_{2DSig} \rangle \triangleright \langle v_{SymS} \rangle \triangleright \langle v_{RoS}, v_{CroS}, v_{TrS}, .. \rangle$ (see Fig. 6.14).

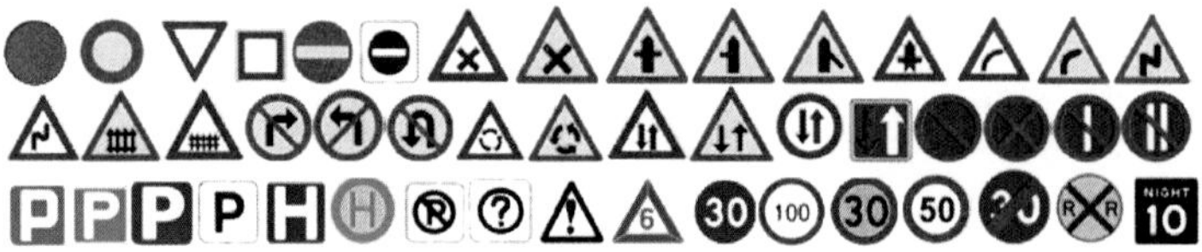

Fig. 6.14. Examples of members of the different categories of road signs

The trademark category refers to a modern trademark that interprets the character of its wearer by associating it with sharply defined signs. Modern trademarks are characteristic symbols of the company. The category of trademarks is divided into the category of editorial trademarks and the category of industrial trademarks and is given by the following categorical chain: $O \triangleright \langle v_{Sig} \rangle \triangleright \langle v_{2DSig} \rangle \triangleright \langle v_{SymS} \rangle \triangleright \langle v_{TrS} \rangle \triangleright \langle v_{PrCo}, v_{Ind}, .. \rangle$. The specific categories of the category of editorial trademarks such as the category of Elsevier trademarks, the category of Prentice-Hall trademarks or the category of Springer-Verlag trademarks are derived from the category of editorial trademarks (see Fig. 6.15).

Fig. 6.15. Examples of modern trademarks

The category of cross signs is based on the existing knowledge of the different types of the cross. The category of crosses is divided into the Latin cross category ν_{Lat}, the Saint Andrew cross category ν_X, the Paty cross category ν_{Pat}, the Papal cross category ν_{Pap}, the Lorraine cross category ν_{Lor}, the Maltese cross category ν_{Mal}, the Chi-Rho cross category ν_{ChR}, or the Celtic (Iona) cross category ν_{Cel} Fig. 6.16 shows examples of members of the different categories of the cross.

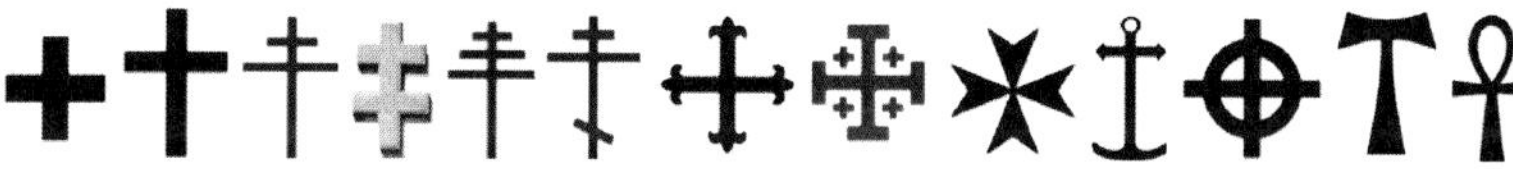

Fig. 6.16. Examples of members of the different categories of the cross

A letter, in this book, denotes any written symbol that is part of the script of any language. Writing is a form of human communication by means of a set of visible marks that are related, by convention, to some particular structural level of language. Language is a system of conventional spoken or written symbols by means of which human beings communicate. Fig. 6.17 shows examples of hieroglyphs, early Sumerian script and cuneiform script. The category of pictograms and the category of hieroglyph are derived from the category of logographic languages.

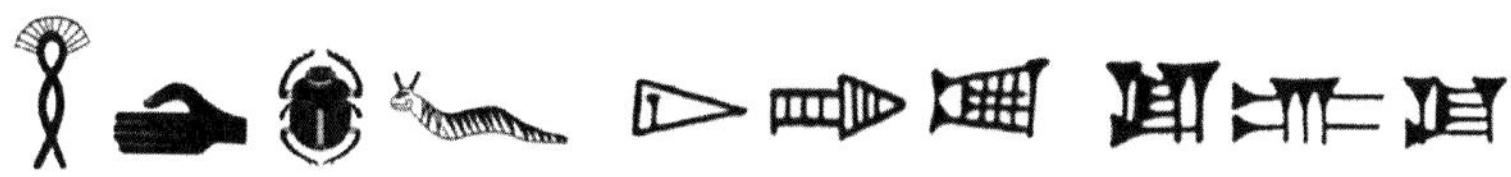

a b c

Fig. 6.17. Examples of: a. hieroglyphs, b. early Sumerian script, c. Cuneiform script

The category of letters is divided into the category of logographic letters ν_{Log}, the category of syllabic letters ν_{Syl}, and the category of alphabetic letters ν_{Alp} and is represented by the categorical chain as follows: $o \triangleright \langle \nu_{Let} \rangle \triangleright \langle \nu_{Log}, \nu_{Syl}, \nu_{Alp} \rangle$. The category of alphabetic letters is divided into the category of Latin letters, the category of Greek letters, the category of the Cyrillic letters, the category of Hebrew letters and the category of Arabic letters and is given by the following categorical chain: $O \triangleright \langle v_{Let} \rangle \triangleright \langle v_{Alp} \rangle \triangleright \langle v_{Lat}, v_{Gre}, v_{Cyr}, v_{Heb}, v_{Ara} .. \rangle$. Examples of members of the specific categories derived from the category of alphabetic letters such as the category of Latin letters (c), the category of Greek letters (e), the category of the Cyrillic letters (d), the category of Hebrew letters (b) and the category of Arabic letters (a) are shown in Fig. 6.18.

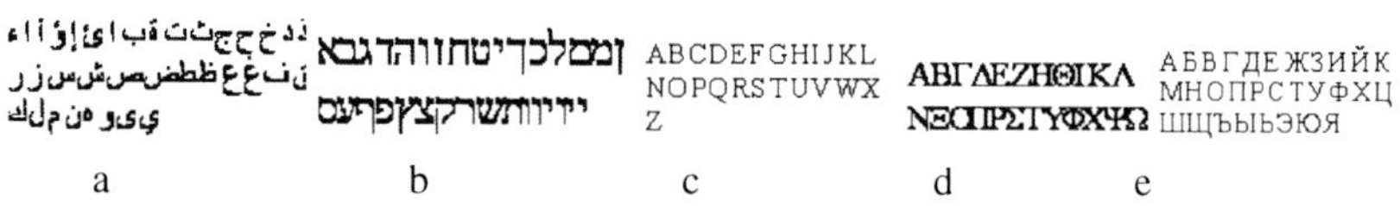

Fig. 6.18. Examples of members of the specific categories derived from the category of alphabetic letters

The letter appearance differs depending on the type of the letter: uppercase or lowercase. In some alphabets the uppercase and lowercase of the same letter look very differently. For this reason the category such as the category of Latin letters is divided into the category of uppercase letters v_{UppC} and the category of lowercase letters v_{LowC} and is given as follows: $O \triangleright \langle v_{Let} \rangle \triangleright \langle v_{Alp} \rangle \triangleright \langle v_{Lat} \rangle \triangleright \langle v_{LowC}, v_{UppC} \rangle$. Fig. 6.19 shows examples of the category of lowercase Greek letters.

α β γ δ ε ζ η θ ι κ λ μ ν ξ ο π
ρ ς σ τ υ φ χ ψ ω

Fig. 6.19. Example of the category of the lowercase Greek letters

The shape of a letter can differ significantly depending on the selected category of font. The category of fonts describes the lowest category called a prototype. The prototype has a well 'defined' shape. For example, the prototype of the letter font such as the Arial font or the Times New Roman font has well defined shapes. The category of fonts such as the category of Latin lowercase letter fonts is divided into the category of the Times New Roman fonts v_{TNR}, the category of the Arial fonts v_{Ar}, and the category of the Bold fonts v_{Bo}. Examples of members of the different categories of different fonts of the letter "T" are shown in Fig. 6.20. In the case of the category of handwritten letters there is a big diversity among shapes of member of the selected specific category e.g. the category of handwritten letter "r".

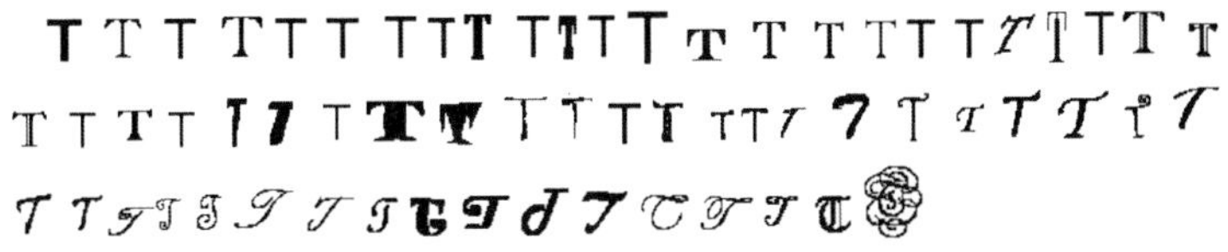

Fig. 6.20. Examples of members of the different categories of different fonts of the letter "T"

A letter is composed into the bigger units such as words, sentences, paragraphs or text. A phrase consists of one or more adjacent words. Phrases have names that reflect the type of word they contain, for example, the noun phrase contains nouns, and the verb phrase contains verbs. The sentence can be dissected into its component phrases, and those phrases into their component words. The analysis of sentences by application of the grammar rules is the task of syntactic analysis. A grammar represents the syntactic rules of the language that are learned as part of the knowledge schema of the knowledge object.

The category of real-world objects refers to the 3D objects that exist in the real world and can be perceived through accessible technical tools such as a camera, a telescope or a microscope. The existing real world objects have different sizes. Based on the size of the objects the category of real world objects is divided into the category of micro-world objects, the category of macro-world objects and the category of earthy-world objects: $O \triangleright \langle V_{\mathrm{Re}O} \rangle \triangleright \langle V_{Mic}, V_{Mac}, V_{Ear} \rangle$.

The category of macro-objects includes objects, size of which is bigger than objects of our today's experience. These objects can be seen only by applying special tools such as a telescope. The category of macro-objects V_{Mac} is divided into the universe category V_{Uni}, the galaxy category V_{Gal}, the star category V_{Str}, the solar system category V_{SolS}, the moon category V_{Mon}, the asteroids category V_{Ast} or the comet category V_{Com} and is given by the following categorical chain: $O \triangleright \langle V_{\mathrm{Re}O} \rangle \triangleright \langle V_{Mac} \rangle \triangleright \langle V_{Uni}, V_{Gal}, V_{Str}, V_{SolS}, V_{Ast}, V_{Mon}, V_{Com}, . \rangle$. Examples of members of the category of macro-object are shown in Fig. 6.21.

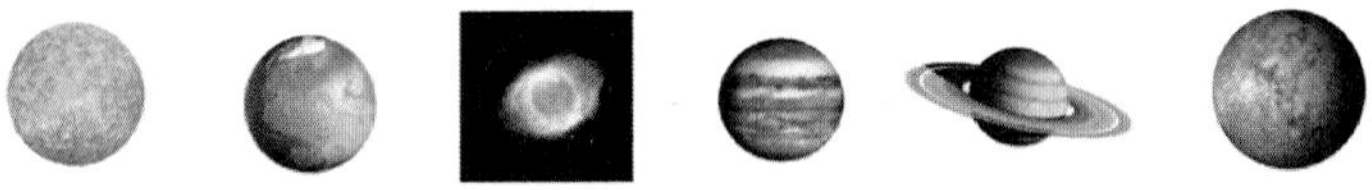

Fig. 6.21. Examples of members of the category of macro-object

Objects that that are too small to be seen by naked eye belong to the category of micro-objects. Members of the category of micro-objects can be registered by applying the one of the methods of photomicrography. The category of micro-world is divided into the category of live and the category of non-live objects. The category of live objects is divided into the category micro-organs V_{Org}, or micro-organisms (bacteria, viruses) V_{Ogn} and is given by the following categorical chain: $O \triangleright \langle V_{\mathrm{Re}O} \rangle \triangleright \langle V_{Mic} \rangle \triangleright \langle V_{Liv} \rangle \triangleright \langle V_{Org}, V_{Ogn} \rangle$. Examples of members of the micro-objects category are shown in Fig. 6.22.

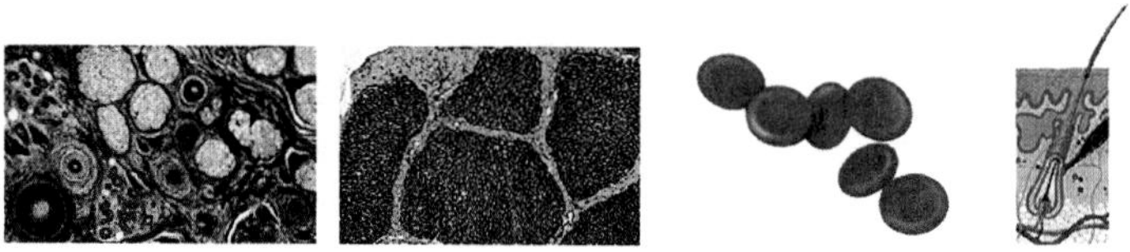

Fig. 6.22. Examples of objects of the micro-world category

The visual objects, size of which is such that can be visible by the naked eye, are objects of the earthy-world objects category. The earthy-world objects category is divided into the category of living objects ν_{Liv} and the category of non-living objects ν_{NLiv} and is given by the following categorical chain: $O \triangleright \langle \nu_{ReO} \rangle \triangleright \langle \nu_{Ert} \rangle \triangleright \langle \nu_{Liv}, \nu_{NLiv} \rangle$. Category of non-living objects ν_{NLiv} includes category of man-made objects ν_{MMan} and the category of non-man-made natural objects ν_{NMan}.

The category of non-living man-made objects covers the broad range of objects that are made for different purposes. The category of man made objects is divided into the category of tools, vehicles, furniture, buildings, arms, or machines: $\langle \Pi \rangle \triangleright \langle \sigma_{Pat} \rangle \triangleright \langle \nu_{ReO} \rangle \triangleright \langle \nu_{Ear} \rangle \triangleright \langle \nu_{NLiv} \rangle \triangleright \langle \nu_{MMad} \rangle \triangleright \langle \nu_{Tol}, \nu_{Veh}, \nu_{Fur}, \nu_{Bul}, \nu_{Arm}, \nu_{Mach}, . \rangle$.

The category of man-made objects can be always broadened about a new model that was lately designed. The category of man-made objects needs to refer to the diversity of objects that are made in the different period of time and the diversity of objects that are results of the new design. For example, furniture of ancient Egypt such as beds are different from today's beds.

The category of mechanical machines is divided into the category of vehicles or the category of non-moving machines. The category of vehicles is divided into the category of air vehicles, the category of space vehicles, the category of water vehicles or the category of land vehicles and is given as: $\langle \Pi \rangle \triangleright \langle \sigma_{El} \rangle \triangleright .. \triangleright \langle \nu_{MMad} \rangle \triangleright \langle \nu_{Veh} \rangle \triangleright \langle \nu_{AirV}, \nu_{SpcV}, \nu_{WatV}, \nu_{LanV}, .. \rangle$. Fig. 6.23 shows examples of members of the different categories of land vehicles. Figs. 6.23 (a-d) show silhouettes of the members of the car category. Fig. 6.23 e, f shows the line-drawing of the different visual aspects of the member of the same car category.

Fig. 6.23. Examples of the different perceptual categories (the silhouette category and the line-drawing category) used to represent members of the ontological category, namely, the vehicle category

A tool is an instrument for making material changes on other objects, as by cutting, shearing, striking, rubbing, grinding, squeezing, measuring, or other process. The specific tools category such as the category of carpenter tools is divided into the hammer category, the chisel category, the saw category, the hook category, the plane category, or the handle category and is given as: $..\triangleright\left\langle v_{\Pr f}\right\rangle\triangleright\left\langle v_{Car}\right\rangle\triangleright\left\langle v_{HamC}, v_{ChisC}, v_{SawC}, v_{PlaC}, v_{HanC}.\right\rangle$. Examples of tools are shown in Fig. 6.24.

Fig. 6.24. Category of gardener tools

The category of living objects v_{Liv} is divided into the category of human beings v_{Hum}, the category of animals v_{Ani}, the category of plants v_{Pla}, the category of fungus v_{Fun}, the category of protoctist $v_{\Pr o}$ and the category of moneran v_{Mon} and is given as: $..\triangleright\left\langle v_{Liv}\right\rangle\triangleright\left\langle v_{Hum}, v_{Ani}, v_{Pla}, v_{Fun}, v_{\Pr o}, v_{Mon}\right\rangle$. The examples of members of the category of mammalian are shown in Fig. 6.25.

Fig. 6.25. Example of members of the category of animals

Within the plant kingdom, plants are divided into two main groups: the plants that produce seeds (flowering plants) and the other group that contains the seedless plants that reproduce by spores. The category of plants is divided into the specific categories based on the botanical taxonomy. The category of real world objects such as plants can have many different perceptual representations. The examples shown in Fig. 6.26 are members of the silhouette category. As it can be seen, even this simple visual representation makes it possible to differentiate among the different trees.

Fig. 6.26. Examples of members of the different categories of tree represented as the silhouette – one of the perceptual categories

The real world object such as a plant consists of different parts. The part category is introduced to represent the different parts of the object. The part category is an auxiliary category that can be derived from any part of the categorical hierarchy. The schema of the part category shows the links to categories that constitute an object. For example, typical flowering plants such as a tree consists of roots, a trunk, stems, leaves, flowers, fruits, and seed. The schema of the part category includes the specific categories that refer to parts of the object. For example, the tree category consists of the different specific categories given by the following schema of the part category: $..\langle v_{Pla} \rangle \triangleright \langle v_{Tre} \rangle \succ [\tau_{Rot}, \tau_{Trn}, \tau_{Stm}, \tau_{Lef}, \tau_{Flw}, \tau_{Frt}, \tau_{Sed}]$. Each part category such as the roots category τ_{Rot}, the trunks category τ_{Trn}, the stems category τ_{Stm}, the leaves category τ_{Lef}, the flowers category τ_{Flw}, the fruits category τ_{Frt}, and the seeds category τ_{Sed} refers to the parts of the tree. Fig. 6.27 shows characteristic parts of the tree. The part can be regarded as an independent object that in turn consists of other parts. For example, a fruit shown in Fig 6.27 consists of characteristic parts such as seeds. Most fruits grow on trees, whereas some of them such as muskmelons grow on creeping vines. The muskmelon has a hard rind that encases the juicy pulp, and flat seeds that form a netlike mass in the hollow centre. Fig 6.27 (a) shows the tangerine, whose fruits are produced from a small, thorny tree that bears simple leaves and orange like blossoms. Fig. 6.27 (b) shows the common orchard fig, a bush-like tree with deeply lobed leaves. Its fruit is a fleshy receptacle (cross section, center) containing numerous small seed. Fig 6.28 shows examples of the category of the parts of a plant.

a b

Fig. 6.27. The example of the category of parts of a tree and the category of parts of a shrub

a b c d e

Fig. 6.28. Examples of the category of the parts of a plant

A fruit is the ripened ovary of any flowering plant, or angiosperm, and usually contains one or more seeds. The knowledge schema for the fruit category includes the visual concept ∂_{ViC}, the name ∂_{Nam}, the definition ∂_{Def} and the method of exemplar generation ∂_{MGe} and is given as follows: $..\triangleright\langle\kappa_{Fru}\rangle\prec\{\partial_{ViC},\partial_{Nam},\partial_{MIn},\partial_{Def},\partial_{MGe}\}$. For example, the definition of the category of fruits includes, among others, the following parts and can be given in the following form: fruit -is-part-of- [plant], fruit-consists of – [blade, core], fruit-is- [outgrowth from the stem of plant]. The definition includes links to other knowledge categories that usually contain the non-visual knowledge. The fruit can be grown on the trees, shrubs or vines. The tree fruits category is divided into the category of plums, apples or pears: $..\langle\nu_{Pla}\rangle\triangleright\langle\nu_{Tre}\rangle\succ[\tau_{Frt}]\triangleright\langle\nu_{Plu},\nu_{App},\nu_{Pea},..\rangle$. The examples of the different tree fruits categories are given in Fig. 6.29 such as the quince category (a), the pear category (b), the lime category (c), the plum category (d), the mango category (e), the papaya category (f), the papaw category (g), the citrus category (h) and the grapefruit category (i).

Fig. 6.29. Examples of members of the tree fruits category

The category of imaginary objects includes the category of objects of scientific visualization ν_{SciV}, the category of objects of literature fiction ν_{InvL}, the category of visual art objects ν_{InvA}, the category of 3D fictious figures ν_{FarT} and the category of mythological objects ν_{Mit}. Dwarfs (dvergar) play a part in Norse mythology. Greek centaurs (in Greek mythology) are part horse and part man, dwelling in the mountains of Thessaly and Arcadia. Fig. 6.30 shows examples of members of the category of imaginary objects: 3D fictious (invented) tree and the category of the mythological objects.

Fig. 6.30. Examples of category of imaginary objects: 3D fictious (invented) tree and the category of the mythological objects

The scientific visualization is focused on generation of the visual objects that are visual representative of the model of the examined phenomenon. The category of scientific visualization is divided into the category of the schematic visualization and the realistic visualization. The category of schematic visualization is divided into the category of diagrams, the category of maps, the category of diagrammatic representations or the category of schematic data visualization. The category of the realistic visualization is divided into the category of modeling of the non-visible phenomenon and the category of realistic data visualization. Fig. 6.31 shows examples of the different categories of scientific visualization that involves the category of schematic visualization and the category of realistic visual representations.

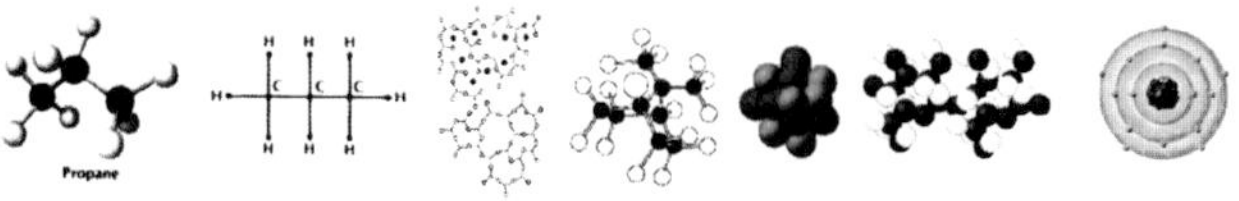

Fig. 6.31. Category of the scientific visualization

The category of the earthy objects is divided into the category of living objects, the category of non-living objects and the category of processes: $O \triangleright \langle v_{\mathrm{Re}O} \rangle \triangleright \langle v_{Ear} \rangle \triangleright \langle v_{Liv}, v_{NLiv}, v_{\mathrm{Pr}oc} \rangle$. The category of processes refers to changes of the visual objects that can be observed during the period of time. The visual process is often represented by the category of animation that is one of the structural categories. The process category is divided into the natural process category and the artificial process category: $.. \triangleright \langle v_{\mathrm{Pr}oc} \rangle \triangleright \langle v_{NatP}, v_{ArtP} \rangle$. The category of natural processes is divided into atmospheric processes, acoustic processes, physical processes, chemical processes, geological processes or biological processes: $.. \triangleright \langle v_{\mathrm{Pr}oc} \rangle \triangleright \langle v_{NatP} \rangle \triangleright \langle v_{AtmP}, v_{AcuP} v_{PhyP}, v_{ChmP}, v_{GeoP}, v_{BioP}, .. \rangle$. The category of physical processes is divided into the category of changing state processes (melting, boiling, freezing), the category of heat transfer processes (convection, conduction, radiation), the category of radioactivity processes, the category of magnetism processes, the category of sound waves processes, or the category of electromagnetic wave processes (light, ultraviolet, infrared). For example, the category of chemical processes is connected with the chemical reactions.

The category of acoustic processes is divided into the category of music, the category of songs or the category of noise: $O \triangleright \langle v_{\mathrm{Re}O} \rangle \triangleright \langle v_{Ear} \rangle \triangleright \langle v_{NLiv} \rangle \triangleright \langle v_{NatP} \rangle \triangleright \langle v_{Acus} \rangle \triangleright \langle v_{Mus}, v_{Son}, v_{Spi}, v_{Noi}, .. \rangle$. The visual representative of the acoustic processes is the sound wave that can be given as the signal in the time domain or transformed into the frequency or time-frequency domain (see Fig. 6.32).

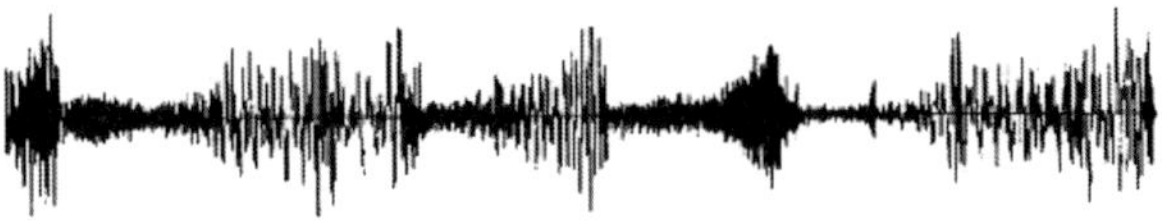

Fig. 6.32. Example of the visual representative of the acoustic process - the sound wave.

The category of geological processes is divided into the category of tectonic processes (earthquake, volcano), the category of erosion processes and the category of sedimentation processes: $..\triangleright\langle v_{\Pr oc}\rangle\triangleright\langle v_{NatP}\rangle\triangleright\langle v_{GeoP}\rangle\triangleright\langle v_{TecP}, v_{Ero}, v_{Sed}, \cdot\rangle$. Fig. 6.33 shows the geological process as a sequence of pictures that are representative of the perceptual animation category.

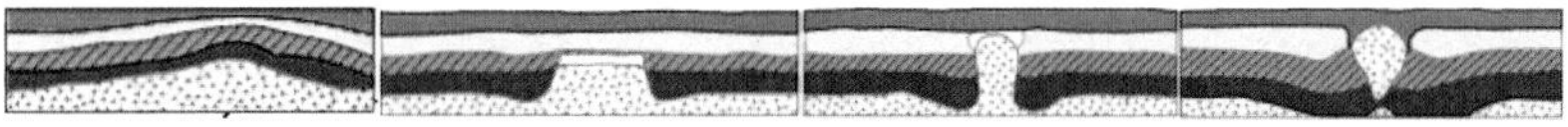

Fig. 6.33. Category of geological process - schematic representation

7 Understanding and Learning of the Knowledge of the Different Categories of Objects

In this Chapter the theoretical framework of the knowledge implementation method is presented. In Chapter 8 the knowledge implementation approach is applied for learning of the knowledge and skills of the different categories of objects. The shape understanding system (SUS) operates in two main modes, the learning mode and understanding mode. Learning and understanding are complementary processes. The ability of SUS to understand depends on the effectiveness of learning process and, in turn, learning of the new knowledge depends on the SUS ability to understand. The knowledge implementation is based on assumption that a system to be able to understand needs to learn and learned knowledge needs to be fully understood, that means, that there is a relation among specific learned facts stored in memory. The knowledge implementation is concerned with learning of the visual and the non-visual knowledge from the different categories of objects. The category of the visual objects was described in Chapter 6. The category of the sensory objects and the category of the text objects are related to the category of the visual objects and these categories will be defined in this Chapter. These categories are described in the context of learning and understanding of the visual object, the sensory object and the text object. Proposed new learning methods, that are part of the knowledge implementation approach, are designed to learn the knowledge of an object from the selected category such as the category of the visual objects, the category of the sensory objects or the category of the text objects. In Section 7.1 understanding and learning of the knowledge of the visual object is presented, in Section 7.2 understanding and learning of the knowledge of the sensory object is presented and in Section 7.3 understanding and learning of the knowledge of the text object is presented. Understanding and learning of the knowledge of the visual object, described in Section 7.1, is considered as acquiring the complex perceptual skills by SUS. Acquiring the complex perceptual skills by SUS, needed in perceiving of the world, is connected with implementation of the sophisticated image processing algorithms and learning of the complex patterns of the visual reasoning sequences. The visual knowledge is learned based on the sample of the visual objects that represent the category of visual objects selected for learning and utilization of the different forms of the visual abstraction. The important forms of the visual abstraction are generalization and specialization described in Sections 7.1.1 and 7.1.2. Learning of the visual knowledge is called the categorical learning and the short description of this method was presented in

Z. Les & M. Les: Shape Understanding System - Knowledge Implementation and Learning, SCI 425, pp. 61–121.
springerlink.com

[1]. In this Section the new developed categorical learning techniques: learning by the small alternation of the visual object (LSA), learning from the simple to complex (LSC), learning by simplification of the complex object (LSCO), learning from parts (LP), and learning parts decomposition (LPD), are described. The new developed categorical learning techniques are applied to learn the knowledge of the selected ontological categories. Learning of the knowledge of the selected ontological categories will be described in Chapter 8.

Understanding and learning of the knowledge of the sensory object, described in Section 7.2, is considered as acquiring the complex perceptual skills by SUS. The category of sensory objects such as the minerals category described in this Chapter is derived from the category of visual objects. An object from the category of visual objects or sensory objects is understood by naming of the examined object. After naming, the object is interpreted based on the learned knowledge. Understanding of the complex visual or sensory object is to know from which parts it consists of, what material is it made of, what for is it used, how to make it, how to use it and how does it work.

Understanding and learning of the knowledge of the text object, described in Section 7.3, is considered as acquiring the complex interpretational skills by SUS [109]. Acquiring the complex interpretational skills by SUS, needed in finding of the meaning of the text is connected with learning of the query-form, the basic-form, the procedural-form and the explanatory and interpretational script. In this book, learning and understanding of the text that belongs to one of the text categories such as the category of the text-query, the category of the text-task, and the category of the dictionary-texts, is presented. The educational tests described in this Chapter, that are used to test the ability of students to understand, are applied to test the understanding abilities of SUS. Solving educational tasks is performed in two main stages, at the first stage SUS is engaging in understanding of the task and at the second stage SUS is solving the task.

7.1 Understanding and Learning of the Knowledge of the Visual Objects

The category of the visual objects was described in Chapter 6. In this Chapter understanding and learning of the knowledge of the visual object is presented. As it was described, understanding and learning of the knowledge of the visual object is considered as acquiring the complex perceptual skills by SUS needed in perceiving of the world. Acquiring of the complex perceptual skills by SUS is connected with implementations of the sophisticated image processing algorithms and learning of the complex patterns of visual reasoning sequences. The visual knowledge is learned based on the sample of the visual objects that represent the category of the visual objects selected for learning and utilization of the different forms of visual abstraction. The important forms of visual abstraction, the generalization and specialization, will be presented in Sections 7.1.1 and 7.1.2. Visual abstraction utilizes the visual abstraction transformations. The simplest visual abstraction transformation transforms the visual object into one of the perceptual categories such as the color-object category, the shaded-object category, the line-drawing category or the silhouette category (see Fig. 7.1). The complex visual abstraction utilizes the visual abstraction transformations that are based on the colour, shape and texture.

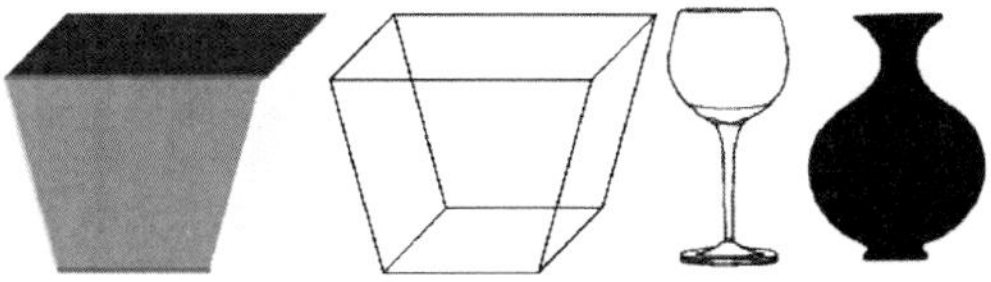

Fig. 7.1. The simplest form of the visual abstraction

The visual abstraction is often applied during thinking and explanatory processes. In order to explain the phenomena or the object, the different forms of abstractions are used. The most important form of the visual abstraction is the schema of the visual object or phenomenon. The schema is showing the selected aspects of the object. Examples of the different types of visual abstractions are shown in Fig. 7.2.

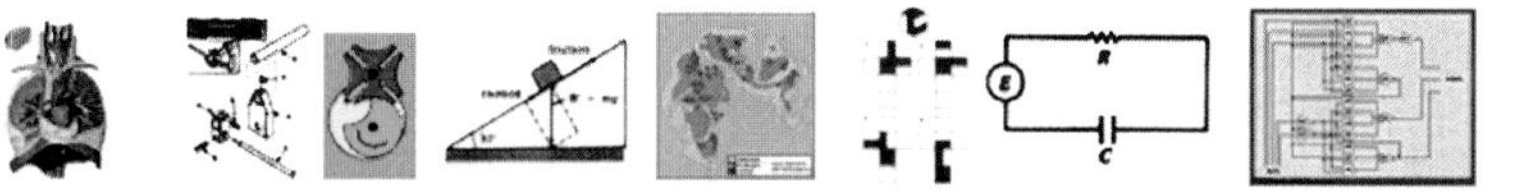

Fig. 7.2. The different forms of visual abstraction

The visual abstraction is based on visual similarity between the visual object and its abstract representation. In the symbolic representation, the visual form of the symbol usually does not bear any kind of similarity to the visual object. Examples of the visual objects and their symbolic representations are shown in Fig. 7.3.

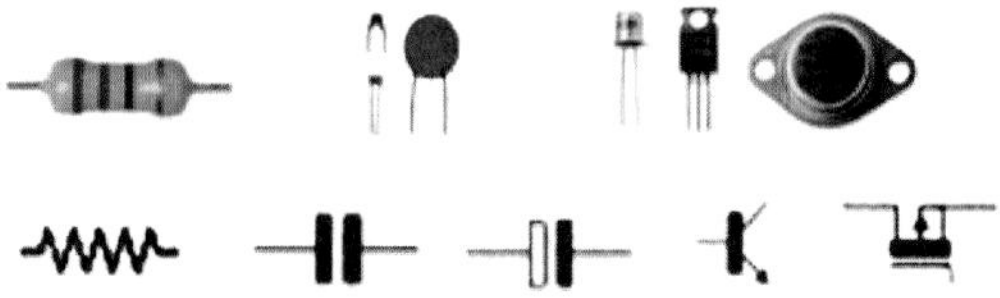

Fig. 7.3. The category of electronic elements and electronic symbols: a resistor, a capacitor, a transistor

In this book the term "learning of the knowledge of the visual object" will be often exchanged by the term "learning of the visual object". In the following Chapters generalization and specialization, as the important forms of visual abstraction, will be presented.

7.1.1 Understanding and Learning of the Knowledge of the Visual Objects – Generalization

Generalization, as one of the important forms of visual abstraction, is often applied during learning and understanding processes. Generalization will be explained based on the examples of objects shown in Fig. 7.4.

Fig. 7.4. Example of generalization

Let's assume that, at first, the knife K1 (Fig 7.4a) is learned and the following symbolic name $\{K\}Q_4^1[M_{1,3}^2[L_T^5]](L_R^3)$ is obtained as the result of learning. Next, the knives K2 (Fig 7.4b) and K3 (Fig 7.4c) are given to SUS for understanding (naming). During naming, the symbolic name $\{K\}Q_4^1[M_{1,3}^2[L^5]](L_R^3)$ is assigned to these two objects (knives). In this example, SUS does not have such a symbolic name stored among the learned visual concepts, then SUS needs to perform generalization. During generalization the specific features of the objects (knives) are "omitted" and the symbolic name is transformed into the simpler form according to the generalization rules. Let's assume that according to the generalization rule (R1) the index in a symbol that represents the generic polygon of the curvilinear class $M_{1,3}^2[L^5]$ is omitted. As the result of applying the generalization transformation (generalization rule), the following symbolic name $\{K\}Q_4^1[M_{1,3}^2[L^5]](L_R^3)$ is obtained. This name is found during the reasoning process as the symbolic name of the visual concept of the object K1. As the result, the objects K2 and K3 will be named "like-knife K1".

Presented example showed that learning of the knowledge of the visual object is connected with learning of generalization rules. The rules of generalization can be defined based on the geometrical properties of the specific visual category. For example, in the case of the symbolic name $\{K\}Q_4^1[M_{1,3}^2[L_T^5]](L_R^3)$ the rules of generalization are given at many different levels of generalization in the form of the symbolic names: (1) $\{K\}Q_4^1[M_{1,3}^2[L^5]](L_R^3)$, (2) $\{K\}Q_4^1[M_{1,3}^2[L^5]](L^3)$, (3) $\{K\}Q_4^1[M_{1,3}^2[L^5]](L)$, (4) $\{K\}Q_4^1[M_{1,3}^2[L^5]](\Lambda)$, (5) $\{K\}Q_4^1[M\ [L^5]](\Lambda)$, (6) $\{K\}Q_4^1[L^5](\Lambda)$, (7) $\{K\}Q_k^1[L\ ](\Lambda)$, (8) $\{K\}Q_k^1[\Lambda](\Lambda)$.

During the next learning stage, in order to differentiate between the two objects K2 and K3, the new knowledge and new skills are learned. These examined objects are different types of knives (type T1 and T2) so there is a need to derive the new classes from the class L^5 to capture visual appearances of these new categories (types). The new classes L^5_B and L^5_b, that are derived from the class L^5, are used to represent knives K2 and K3 by the symbolic names as follows: $\{K\}Q^1_4[M^2_{1,3}[L^5_b]](L^3_R)$ (K2 type T2) and $\{K\}Q^1_4[M^2_{1,3}[L^5_B]](L^3_R)$ (K3 type T3). These symbolic names are stored as part of the visual concepts of knives K2 and K3.

Let's assume that during naming of the object K4 (type T4) (see Fig. 7.4d), the symbolic name $\{K\}Q^1_3[M^2_{0,3}[L^4]](L^3_R)$ is obtained. Similarly like in the previous example, during generalization process the name $\{K\}Q^1[M^2[L\ ]](L^3_R)$ is assigned to the examined object and the object is named "like-knife". Because this object is the member of the knife category T4 (type T4), during the learning stage the new class L^4_b is derived from the class L^4. The class L^4_b is the class associated with the class L^5_b and it can be seen as the class for which the side (2) of the archetype of the class L^5_b has the length l=0. The new class L^4_b, that is derived from classes L^5_b and L^4, are used to represent the knife K4 by the symbolic name $\{K\}Q^1_4[M^2_{1,3}[L^4_b]](L^3_R)$ and this symbolic name is stored as part of the visual concept of the knife K4.

Objects, for which symbolic names are given by associated classes, are very similar and during the understanding process more than one symbolic name, that comes from these associated classes, can be assigned to the examined object. Classes L^5_T, L^5_B, L^5_b and L^4_b described in this section are associated classes. Associated classes are written as $\{L^5_T, L^5_B\}$, $\{L^5_B, L^5_b\}$, $\{L^5_b, L^4_b\}$. Fig 7.5 shows examples of associated classes L^5_T, L^5_B, L^5_b and L^4_b.

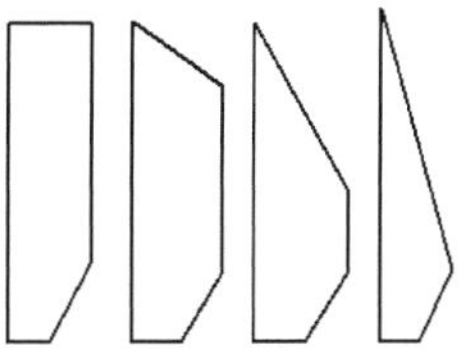

Fig. 7.5. Associated classes

A visual name is represented in a general form and is utilized during a generalization process. For example, the objects from the knife category shown in Fig.7.6 can be represented by the general form of the symbolic name $\{K\}Q^1_4[X](Y)$, where $X = \{L^5_T, M^1_{ia}[L^5_T], M^2_{1a,3b}[L^5_T]\}$ and

$Y=\{L^3,M^1_{ia}[L^3],M^2_{1,3}[L^3],Q^1_i[L^3](M^1),L^4_{\{A,W\}},M^1_{ia}[L^4_{\{A,W\}}],M^2_{i,j}[L^4_{\{A,W\}}]\}$. Generalization rules can be given in the general form of the symbolic names. During learning process, after learning of the prototypes of the knife category, the learned visual knowledge can be represented by using the general form of the symbolic name: $\{K\}Q^1_4[M^2_{1a,3b}[L^5_T]](Y)$. Learning of the new objects from the knife category will specify the class Y. Let's assume that the learned classes are given as $Y=\{L^3,M^1_{ia}[L^3],M^2_{1,3}[L^3]\}$. Each symbolic name refers to the different types of knife (Ta, Tb, Tc). In the case when a new object T4 is learned, its symbolic name $\{K\}Q^1_4[M^2_{1a,3b}[L^5_T]](M^1_{ia}[L^3_O])$ is compared and the term $M^1_{ia}[L^3_O]$ is added to the visual concept represented by the general form of the symbolic name $Y=\{L^3,M^1_{ia}[L^3],M^2_{1,3}[L^3],M^1_{ia}[L^3_O]\}$. In the case when an object perceived during the recognition stage has the symbolic name $\{K\}Q^1_4[M^2_{1a,3b}[L^5_T]](Q^1_i[L^3](M^1))$, SUS recognizes it as the knife-like object, naming it T-knife-like and the part $Q^1_i[L^3](M^1)$ is added to the learned visual concept $Y=\{L^3,M^1_{ia}[L^3],M^2_{1,3}[L^3],M^1_{ia}[L^3_O],Q^1_i[L^3](M^1)\}$.

Fig. 7.6. Examples of learned objects from the knife category

In the case of symmetrical objects that are derived from the same generic class there is a need to derive the specific classes that make it possible to see differences among very visually similar objects and to see differences among objects of the different categories. For example, objects shown in Fig. 7.7 are all from the class $\{E\}Q^2[V](XY)$, where $V=M^1_{3t}[L^4_R]$, $X\equiv\Lambda$, $Y\equiv\Lambda$ and symbol $\bullet$ denotes that the sizes of both residues are equal. In order to name it properly, the additional specific classes need to be derived. The object O1 shown in Fig. 7.7.a and the object O2 shown in Fig. 7.7.b looks similar although the sizes of the residues are different. The symbolic name for the object O1 is $\{E\}Q^2[V](2\bullet X)$, and for the object O2 is $\{E\}Q^2[V](2X)$ and $X\equiv L^3$. The object O3 showed Fig. 7.7.c looks to be very different from both objects O1 and O2. In order to capture these visual differences the new class is derived. The symbolic name for the object O1 is $\{E\}Q^2_{2,4}[V](2\bullet X)$, for the object O2 is $\{E\}Q^2_{2,4}[V](2X)$ and for the object O3 is $\{E\}Q^2_{1,2}[V](2X)$. The symbolic name for the object O4 shown in Fig. 7.7.d is given as $\{E\}Q^2_{2u,4u}[V](2\bullet X)$, whereas the object O5 shown in Fig. 7.7.e is given as $\{E\}Q^2_{2m,4m}[V](2X)$. The object O5 looks very different from the objects O4 and O6, shown in Fig. 7.7.f, partially because this object is not symmetrical due to two different types of placement of residues. The object O7 (Fig. 7.7.g), described by the symbolic name $\{E\}Q^2_{2m,4m}[V](2*X)$, has the same residuals placed in the same way but their orientation is different what is marked by the symbol "$*$". The object O8 (Fig. 7.7.h) described by the symbolic name $\{E\}Q^2_{2m,4m}[V](2\circ X)$, similarly like the object O7, has the same residuals placed in the same way but their sizes are different what is marked by the symbol "$\circ$". The object O9 (Fig. 7.7.i), described by the symbolic name $\{E\}Q^2_{2u,4m}[V](2X)$, has the same residuals but placed in the different way. The objects (O10-O13), $\{E\}Q^2_{2m,4m}[U](2\bullet X)$ have different residuals than the objects (O1-O9) $\{E\}Q^2_{2m,4m}[U](2\bullet X)$, where $X\in\{M^1[L^4_T],M^1_C,Q^2[L^3][2L^3]\}$. The object O9 $\{E\}Q^2_{2m,4m}[V](XY)$ has two different residuals L^3_R, and L^4_R.

Fig. 7.7. Examples of dissimilar objects

The visual differences can be due to properties of the curvilinear segments. In order to depict the visual differences of the different objects the new specific curvilinear classes need to be derived. These classes are described in terms of the differences in the type of the curvilinear segments. The type of the curvilinear segment is denoted by the variable t, domain of which is the type of curvilinear segment $t\in[a,b,c,d,...]$, where letters $a,b,c,d,...$ denotes the type of the curvilinear segment. Fig. 7.8 shows objects $\{E\}Q^2_{2m,4m}[M^1_{3t}[L^4_R]](2\bullet Y)$ that have the different curvilinear segments.

Fig. 7.8. Examples of similar objects

7.1.2 Understanding and Learning of the Knowledge of the Visual Objects – Specialization

Specialization, that is the important form of the visual abstraction, is often applied during learning and understanding process. Specialization is connected with learning and understanding of the similar objects. Learning of the visual concept of the ontological category is connected with solving the problem of discrimination among similar objects (visual similarities). Similarity of objects can be regarded at the two levels: the conceptual similarity and the visual similarity level. Conceptual similarity of objects came from belonging to the same category e.g. tools for eating (a fork, a knife). Visual similarity is concerned with visual similarity of the visual objects (phantoms). Knowledge about objects that are members of the ontological category needs to be learned in the context of similar objects. Similar objects are objects that share some of the visual features. Learned visual concepts of the ontological category include the symbolic names of the most similar objects. During naming (understanding) these similar objects can be interpreted as the objects from the same category v^i , in the situations when the perceived object is not well visible and only the contextual information indicates that the perceived object can be the object of the category v^i .

During learning of the visual knowledge of objects that are members of a given category, there is a need to learn of the visual knowledge about objects that look very similar. In order to be able to discriminate among similar objects, SUS needs to learn new skills and new knowledge. Learning new skills and new knowledge is connected with learning of the new specific classes and implementing the new processing methods. In the case of objects with concavities the proper discrimination among objects requires to derive the new specific classes based on many different properties of the concave objects. For example, the object shown in Fig. 7.9a (from the class $Q^2[L_R^4](2L_R^4)$) and Fig. 7.9b (from the class $Q^3[L_R^4](3L_R^4)$) can be easy to recognize by counting a number of residuals. However, all other objects shown in Fig. 7.9 are objects from the class $Q^4[L_R^4](4L_R^4)$. For description of the meaning of the symbolic names there is a need to refer to our first book [1]. In order to recognize objects shown in Fig. 7.9 (c-h) there is a need to derived specific classes. The class that represents all objects shown in Fig. 7.9, is denoted as $Q^k[L_R^4](kX)$, and is derived from the class L_R^4 . The specific class that can capture the differences among objects is given as $Q^k_{i(h)}[L_R^4](kX)$, where i is a number of sides with residuals and h is a number of residuals at each side. Based on the symbolic names of the newly derived specific classes, each learned object now has the unique symbolic name that makes it possible to recognize each object shown in Fig. 7.9(c-h). During learning process each object shown in Fig. 7.9 is at first assigned to the object category B (Fig.7.9a), C (Fig.7.9b), A1-A6 (Fig.7.9c-h), and next the symbolic name is assigned to each object, denoted as follows: $B \triangleright Q^4_{1,2}[L^4_R](4L^4_R)$, $C \triangleright Q^4_{1,2,3}[L^4_R](4L^4_R)$ $A1 \triangleright Q^4_{1,2,3,4}[L^4_R](4L^4_R)$, $A2 \triangleright Q^4_{1,2(2),3}[L^4_R](4L^4_R)$, $A3 \triangleright Q^4_{1,2,3(2)}[L^4_R](4L^4_R)$, $A4 \triangleright Q^4_{1(2),3(2)}[L^4_R](4L^4_R)$, $A5 \triangleright Q^4_{1,3(3)}[L^4_R](4L^4_R)$, $A6 \triangleright Q^4_{1(4)}[L^R_4](4L^R_4)$.

Fig. 7.9. Learning similar objects by applying specialization (derivation of the new specific classes) (a) $Q^2_{1,2}[L^4_R](2L^4_R)$ (b) $Q^3_{1,2,3}[L^4_R](3L^4_R)$ (c) $Q^4_{1,2,3,4}[L^4_R](4L^4_R)$, (d) $Q^4_{1,2(2),3}[L^4_R](4L^4_R)$ (e) $Q^4_{1,2,3(2)}[L^4_R](4L^4_R)$ (f) $Q^4_{1(2),3(2)}[L^4_R](4L^4_R)$ (g) $Q^4_{1,3(3)}[L^4_R](4L^4_R)$ (h) $Q^4_{1(4)}[L^R_4](4L^R_4)$)

For specific classes Ω_j that were established, there is a need to implement the proper image transformations $\Theta_i(\Omega_j)$ and to learn the reasoning process. Learning of the image transformations $\Theta_i(\Omega_j)$ is connected with learning of the new visual skills. Learning of the new visual skills will be described in Chapter 8.

7.1.3 Understanding and Learning of the Knowledge of the Visual Objects – Categorical Learning

In previous Section generalization and specialization, as the important forms of visual abstraction, were described. In this Section the new developed categorical learning techniques that utilize the different forms of visual abstraction are presented. Learning of the knowledge of the selected categories of visual objects, such as the category of signs, requires learning both the visual and the non-visual knowledge. Learning of the non-visual knowledge is to learn knowledge of the categorical chain and knowledge of the knowledge schema (the name ∂_{Nam}, the definition ∂_{Def}). Learning of the visual knowledge is to learn the visual description κ_c, the symbolic name η_c, the visual concept φ_c, the rules of part decomposition ρ^P_c, the rules of generalization ρ^G_c and the image transformations Θ^i_c. In order to learn of the visual knowledge for the broad spectrum of ontological categories there is a need to elaborate new methods of learning of the new visual knowledge. Learning of the visual knowledge is called the categorical learning and the short description of this method was presented in [1]. In this Chapter the new developed categorical learning techniques are presented. Description of these techniques is focused on presentation of learning algorithms and description of the selection of the proper learning sample that is related to the specific categories of the visual objects. Examples of application of these learning methods for learning of the specific ontological categories such as the category of signs or category of real world object will be presented in Chapter 8.

The visual knowledge of the visual objects is learned using one of the proposed categorical learning techniques:

- learning by the small alternation of the visual object (LSA),
- learning from the simple to complex (LSC),
- learning by simplification of a complex object (LSCO),
- learning from parts (LP), and
- learning parts decomposition (LPD).

7.1.3.1 Learning by the Small Alternation of the Visual Object (LSA)

Learning by the small alternation of the visual object (LSA) is to select the visual object and next, by alternating its visual features, learning of the visual knowledge from a sequence of generated objects. For each selected object, the visual description of the object is learned and, at the end, the symbolic name of the object is learned. The alternated object, that is considered to be the representative of some of the ontological category, is named by the name of that category. Learning by the small alternation of the visual object is used in order to avoid an error connected with 'confusing' naming of objects that are very similar. Learning involves establishing the new specific classes Ω^j that make it possible to assign an object to the very specific shape class and to elaborate the specific image transformations $\Theta^k(\Omega^j)$ in order to process objects from the newly established specific classes. The specific classes that have different symbolic names enable us to name very similar visual objects that belong to the very different ontological categories.

The learning process can be described as follows. For an object o_i^1 from the category c_h the feature f_k is selected, its value is established $f_k = A$ and the new object o_i^2 is generated. If a generated object is the object from the category c_h its name is not changed. If a generated object is an object from the category c_l the name n_l of the category c_l is assigned to the object. This process is represented symbolically as follows: o_i^2: $o_i^2 \equiv c_l \rightarrow o_i^2 \triangleright n_l$. The process of generating of a sequence of objects can be described as: $o_i^1(f_k = A) \rightarrow o_i^2(f_k = B) ... \rightarrow o_i^m(f_k = X) ... \rightarrow o_i^M(f_k = Y)$. Fig. 7.10 shows objects that are generated by alternating features of the object (triangle). At first the one, two, and three segments are introduced (the first row in Fig. 7.10) and more than three segments are added (rows 2 and 3 in Fig. 7.10). In the rows 4 to 6 (in Fig. 7.10) objects that were generated by alternating features of the object (triangles with one hole and convex objects with n-holes) are shown.

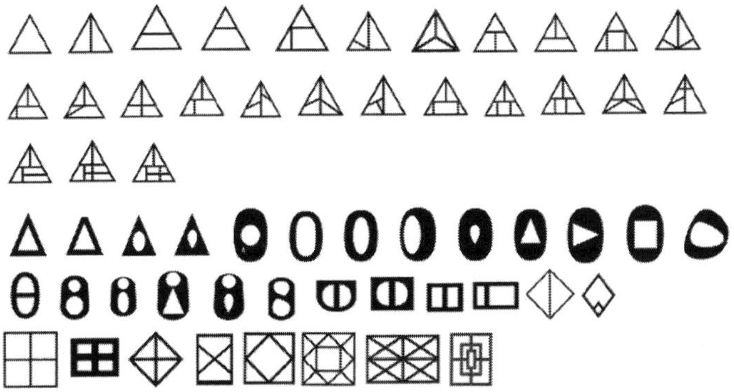

Fig. 7.10. Examples of objects that are generated by alternating features of the selected object

7.1.3.2 Learning from the Simple to Complex (LSC)

Learning from the simple to complex (LSC) is applied when the object to be learned is too complex and there is no well elaborated method of processing (image transformations) for this type of the object. This method can be regarded as a special case of learning by the small alternation of the visual object (LSA). The difference is such that in learning from the simple to complex the alternation of the object is aimed to generate the object that is more complex than the previous one. Increasing the complexity of the object means increasing a number of parts, holes or concavities in the object.

During learning from the simple to complex (LSC), the complex object o_N that is selected for learning is simplified and the sequence of objects $o_1,\ldots,o_N$ is generated. The first object o_1 in the sequence is the simplest version of the object o_N. The object o_1 in a sequence is represented by its characteristic aspect that is more complex than the object o_1 and much simple than an object o_N. During learning process new classes are established and new image transformations are implemented. The symbolic name of the object o_N is learned as the result of derivation of new specific shape classes and implementation of the new image transformations.

Fig. 7.11 shows objects used for learning from the simple to complex. The complex object shown in Fig. 7.11 (e) is selected and the sequence of simplified objects shown in Fig 7.11 (a-d) is generated.

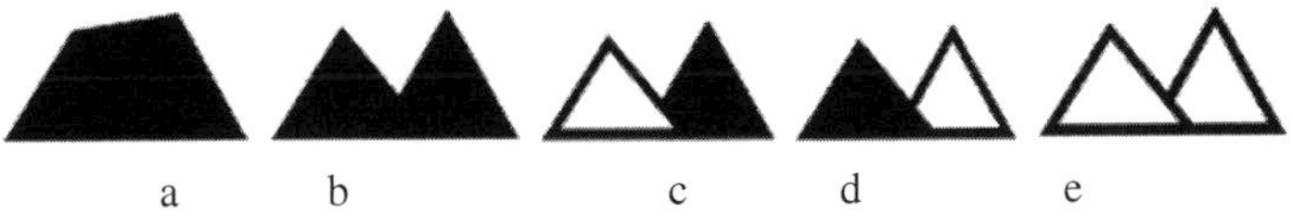

Fig. 7.11. Learning from the simple to complex

Fig. 7.12 shows examples of objects that are generated as the result of simplification of the object , from one of the knife categories.

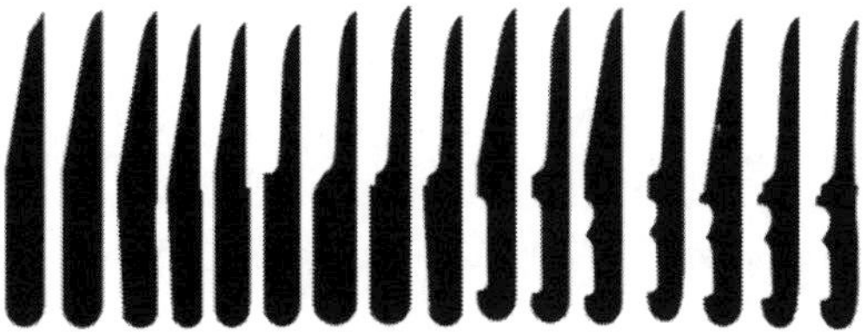

Fig. 7.12. Learning objects from the knife category using the LSC method

Learning from the simple to complex method (LSC) is applied to learn of the signs and real world objects categories described in Chapter 8.

7.1.3.3 Learning by Simplification of a Complex Object (LSCO)

Learning by simplification of a complex object (LSCO) is used at the first stage of learning. The LSCO method is used to learn generalization rules that are stored as part of the knowledge of a reasoning process. Learning by simplification of a complex object (LSCO) is similar to learning from the simple to complex. The main difference in LSC and LSCO learning methods is that during LSC learning, a sequence of objects $o_1,...,o_N$ is generated and for each object o_i from the sequence of objects the symbolic name $\eta(o_i)$ is learned by applying image transformations to the object o_i. When using the LSCO learning method, at first the symbolic name $\eta(o_N)$ for the complex object o_N is learned by applying image transformations to the object o_N. Next, the sequence of symbolic names $\eta_N,...,\eta_1$ is generated by applying of the generalization rules. The generalization rules are based on string transformations such as cutting selected parts. Each symbolic name η_i represents the object o_i that is generated as the result of application of the image transformation to the object o_{i-1}. Application of the generalization rules by transforming of the string is performed in parallel to application of image transformations such as polygonization (removing curved segments), filling the holes or filling residuals (concavities) to visual object. For example, the transformation of the string parts, called the cyclic parts is performed in parallel to the application of the image transformation of filling holes. During learning of the visual knowledge by using LSCO there is no need to learn the symbolic names from all visual objects. We can construct a symbolic name from the previously learned symbolic names by transforming them according to generalization rules. For example, there is no need to learn a symbolic name for the object O2  if the symbolic name was previously learned for the object O1 . The symbolic name for the object O2 can be obtained by transforming the symbolic description or the symbolic name of this object by applying the generalization rule "cutting acyclic part". Applying the rule "cutting acyclic part" means to apply to the visual object O1, at the same time, the image transformation "filling holes". For example, the learned symbolic description for the object O1 (shown in Fig. 7.13.a) is as follows: A2_Q4_Q1_u_Q1_u_Q1_u_Q1_u_M2_L8_Msn_1qaqaqqq_0k000k00_L3_AE_qaq_M1_L3_AE_qqa_M1_L3_AE_qqa_M1_L3_AE_qaq_M1_K1_K1".

It is relatively easy to obtain the symbolic description for the object O2 (shown in Fig. 7.13.d) by transforming of the string by applying the generalization rule "cutting acyclic part". As the result of string transformation the following symbolic name was obtained: Q4_Q1_u_Q1_u_Q1_u_Q1_u_M2_L8_Msn_1qaqaqqq_0k000k00_L3_AE_qaq_M1_L3_AE_qqa_M1_L3_AE_qqa_M1_L3_AE_qaq_M1.

Learning by simplification LSCO can be described as a set of consecutive steps. At first, the object shown in Fig 7.13.a is selected, the symbolic description is obtained (shown in yellow below) and the symbolic name is extracted (shown in grey below).

1. "A2_Q1_u_L4_Wgn_qqqa_L3_A_L4_Wgn_L3_A"
"[A2][[Q1][L4|Wgn|][|S88||B44,59,100,53||A110,61,58,131||E2||r3||G168||@2210||q3|]{%[|L3|A|]|S38||B84,100,91||A59,52,69||G112||@528|]%}}]{[L4|Wgn|][|S39||B27,75,100,69||A115,58,61,126||E1||r4||G121||@611|]}{[|L3|A|]|S43||B100,81,89||A58,72,50||G123||@645|]}"
2. "A1_Q1_u_L4_Wgn_qqqa_L3_A_L3_A"
"[A1][[Q1][L4|Wgn|][|S88||B44,59,100,53||A110,61,58,131||E2||r3||G168||@2210||q3|]{%[|L3|A|]|S38||B84,100,91||A59,52,69||G112||@528|]%}}]{[|L3|A|]|S43||B100,81,89||A58,72,50||G123||@645|]}".
3. "A1_Q1_u_L4_Wgn_qqqa_L3_A_L4_Wgn"
[A1][[Q1][L4|Wgn|][|S88||B44,59,100,53||A110,61,58,131||E2||r3||G168||@2210||q3|]{%[|L3|A|]|S38||B84,100,91||A59,52,69||G112||@528|]%}}]{[L4|Wgn|][|S39||B27,75,100,69||A115,58,61,126||E1||r4||G121||@611|]}
4. "Q1_u_L4_Wgn_qqqa_L3_A"
„[Q1][L4|Wgn|][|S88||B44,59,100,53||A110,61,58,131||E2||r3||G168||@2210||q3|]{%[|L3|A|]|S38||B84,100,91||A59,52,69||G112||@528|]%}}"
5. "L4_Wgn_qqqa"
"[L4|Wgn|][|S88||B44,59,100,53||A110,61,58,131||E2||r3||G168||@2210||q3|]".

Next, the generalization transformation (cutting string) is performed on the symbolic description and for all combinations of the transformed string the symbolic name is extracted. The simplified cases (visual objects) are generated and tested, as shown in Fig 7.13. (b-e). The specialization by applying transformation on strings (symbolic names) can be applied to generate objects shown in Fig 7.13. (f-g) The transformation performed on the symbolic name Q1_u_L4_Wgn_qqqa_L3_A adds the acyclic parts A1 _K1 or parts A2_K1_K1 that generates acyclic objects shown in Fig 7.13. (f-g). The symbolic names obtained by adding the acyclic parts are as follows "A1_Q1_u_L4_Wgn_qqqa_L3_K1" and "A2_Q1_u_L4_Wgn_qqqa_L3_K1_K1".

Fig. 7.13. Examples of object used in the experiment learning by simplification of the complex object

In Fig. 7.14 (first object at the first row) a more complex object is shown and examples of objects that are generated by applying the generalization transformation (cutting string) to the string (the symbolic description) are shown. These objects were used to learn the visual knowledge during learning by simplification of the complex object. The symbolic description of the complex object (Fig. 7.14 top-left) is as follows:

"[A9][[L4|R|][|S87||B81,100,81,100||A90,90,90,90||E1||r0||G355||@6238|]]{[Q1][|
L5T|]|S35||B32,81,32,76,100||A123,147,90,90,90|E2|r0||G75||@593||q1|]{%[|L3|R|
]|S28||B55,84,100||A90,33,57||G57||@189|]%}}}{[Q1][|L5T|]|S35||B32,100,78,32,
82||A90,90,90,146,124|E2|r3||G75||@593||q2|]{%[|L3|R|]|S28||B100,83,55||A34,90,
56||G58||@192|]%}}}{[Q1][|L5|Mr|]|S18||B26,54,100,26,79||A90,90,90,110,160|E
3|r3||G32||@123||q1|]{%[|L3|R|]|S14||B36,94,100||A89,21,70||G20||@35|]%}}}{[Q
1][|L5T|]|S18||B24,82,27,100,57||A158,112,90,90,90|E3|r0||G31||@122||q2|]{%[|L
3|R|]|S14||B100,93,36||A21,90,68||G22||@39|]%}}}{[L4|R|][|S22||B28,100,29,100|
|A90,90,90,90||E3||r0||G31||@151|]}{[Q1][|L5T|]|S35||B32,82,33,100,78||A147,12
3,90,90,89|E2|r0||G76||@600||q0|]{%[|L3|R|]|S28||B83,55,100||A90,56,34||G58||@
190|]%}}}{[Q1][|L5|Mr|]|S18||B22,100,57,26,81||A90,90,89,157,115|E3|r4||G29||
@112||q0|]{%[|L3|R|]|S14||B91,42,100||A90,65,25||G24||@42|]%}}}{[Q1][|L5T|]|
S18||B23,83,24,55,100||A114,155,91,90,90|E3|r2||G28||@108||q3|]{%[|L3|R|]|S15||
B91,100,40||A24,66,90||G23||@43|]%}}}{[Q1][|L5T|]|S35||B32,77,100,33,81||A91
,90,90,123,146|E2|r3||G76||@597||q3|]{%[|L3|R|]|S28||B100,55,83||A57,90,33||G57
||@187|]%}}}}"

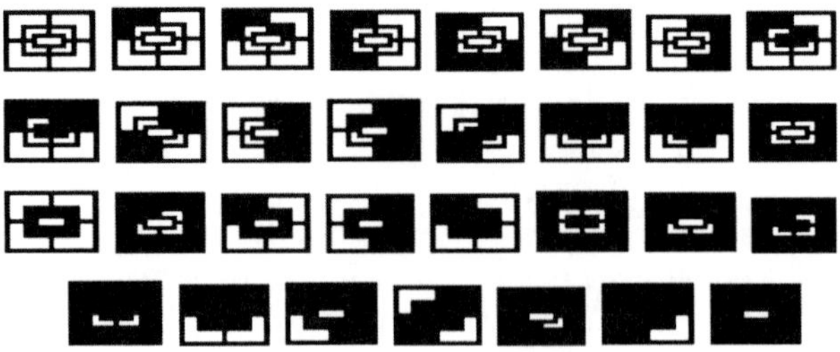

Fig. 7.14. Examples of objects generated during learning by simplification of the complex object

A few examples of the strings obtained during generalization process are presented below:

```
"[A4][[L4|R|][|S87||B81,100,81,100||A90,90,90,90||E1||r0||G355||@6238|]]
{[L4|R|][|S22||B28,100,29,100||A90,90,90,90||E3||r0||G31||@151|]}
{[Q1][|L5|Mr|]|S18||B22,100,57,26,81||A90,90,89,157,115|E3|r4||G29||@112||q0|]{
%[|L3|R|]|S14||B91,42,100||A90,65,25||G24||@42|]%}}}{[Q1][|L5T|]|S18||B23,83,
24,55,100||A114,155,91,90,90|E3|r2||G28||@108||q3|]{%[|L3|R|]|S15||B91,100,40||
A24,66,90||G23||@43|]%}}}{[Q1][|L5T|]|S35||B32,77,100,33,81||A91,90,90,123,1
46|E2|r3||G76||@597||q3|]{%[|L3|R|]|S28||B100,55,83||A57,90,33||G57||@187|]%}
}}}
```

```
"[A3][[L4|R|][|S87||B81,100,81,100||A90,90,90,90||E1||r0||G355||@6238|]]{[L4|R|]
[|S22||B28,100,29,100||A90,90,90,90||E3||r0||G31||@151|]}{[Q1][|L5|Mr|]|S18||B2
2,100,57,26,81||A90,90,89,157,115|E3|r4||G29||@112||q0|]{%[|L3|R|]|S14||B91,42,
100||A90,65,25||G24||@42|]%}}}{[Q1][|L5T|]|S18||B23,83,24,55,100||A114,155,9
1,90,90|E3|r2||G28||@108||q3|]{%[|L3|R|]|S15||B91,100,40||A24,66,90||G23||@43|]
%}}}
```

```
"[A2][[L4|R|][|S87||B81,100,81,100||A90,90,90,90||E1||r0||G355||@6238|]]{[L4|R|]
[|S22||B28,100,29,100||A90,90,90,90||E3||r0||G31||@151|]}{[Q1][|L5|Mr|]|S18||B2
2,100,57,26,81||A90,90,89,157,115|E3|r4||G29||@112||q0|]{%[|L3|R|]|S14||B91,42,
100||A90,65,25||G24||@42|]%}}}
```

```
"[A1][[L4|R|][|S87||B81,100,81,100||A90,90,90,90||E1||r0||G355||@6238|]]{[L4|R|]
[|S22||B28,100,29,100||A90,90,90,90||E3||r0||G31||@151|]}
```

7.1.3.4 Learning from Parts

Learning from parts (LP) is to form the symbolic name of the learned object from the previously learned symbolic names of objects, called parts. The result of learning of the visual object is the symbolic name that is stored as part of the visual concept. Symbolic names $\eta_c^1,...,\eta_c^K$ that were obtained during learning process can be used to form the new symbolic name by transforming strings according to selected generalization rules. During learning from parts (LP) at first, the symbolic names $\eta_c^1,...,\eta_c^K$ are grouped into groups that represent parts of the object used for learning. The object o^i that consists of K parts $p_1^i,...,p_K^i$ is represented as $o^i \triangleright [p_1^i,...,p_K^i]$. The symbolic name that represents the object o^i is denoted as $\eta(o^i)$, and the symbolic name of its j-part is denoted as $\eta(p_j^i)$. The symbolic name of the object o^i, that consists of two parts p_1^i and p_2^i, can be obtained by applying the string combining transformations $\Theta : \eta(o^i) \equiv \eta(p_1^i) \oplus \eta(p_2^i)$, where $\oplus$ is an operator that transforms strings according to the selected generalization rule. During learning the visual knowledge of the object that consists of two parts, at first, from a set of learned symbolic names $\eta_c^1,...,\eta_c^K$ two groups of objects, that represents two parts, are selected $\eta_I^1,...,\eta_I^M$, $\eta_{II}^1,...,\eta_{II}^M$ and next the string η is obtained by applying the string combining transformation $\Theta:\eta \equiv \eta_I^i \oplus \eta_{II}^i$. The obtained symbolic name η represents object o^i so the string combining transformation can be written as $\Theta:\eta(o^i) \equiv \eta_I^i(p_1^i) \oplus \eta_{II}^i(p_2^i)$ or in a general form as $\Theta:\eta(o^i) \equiv \eta\ (p_1^i) \oplus \eta\ (p_2^i)$. Similarly, objects that consist of more than two parts can be obtained by applying the string combining transformation to previously learned strings $\eta_c^1,...,\eta_c^K$.

For example, for an object  the following symbolic name was obtained:

1:A4_Q4_Q1_u_u_u_Q1_u_M1_L6_Rsr_aqa1q1_0000k0_L3_R_qaq_M1_L3_R_L3_R_L3_R_qqa_M1_M1_L3_RE_00k_M1_L3_RE_00k_M1_L3_RE_0k0_M1_L3_RE_00k",

whereas for an object  the following symbolic name was obtained:

2:"A2_Q4_Q1_u_u_u_Q1_u_M1_L6_Rsr_aqa1q1_0000k0_L4_H_qqaq_M1_L3_R_L3_R_L4_Ha_qqaq_M1_M1_L3_RE_00k_M1_L3_RE_00k"

and for an object 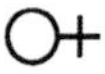 the symbolic name obtained was as follows:

3:A1_Q4_Q2_u_u_u_u_Q2_u_u_M1_L6_aaqaaq_00000k_L4_Wgn_qaqa_L3_R_M1_L3_RE_L3_RE_L4_Wgn_aqaq_L3_R_M1_K1".

In order to learn from parts, at first the main parts are extracted

1:Q4_Q1_u_u_u_Q1_u_M1_L6_Rsr_aqa1q1_0000k0_L3_R_qaq_M1_L3_R_L3_R_L3_R
2:Q4_Q1_u_u_u_Q1_u_M1_L6_Rsr_aqa1q1_0000k0_L4_H_qqaq_M1_L3_R_L3_R_L4_Ha_qqaq_M1
3:Q4_Q2_u_u_u_u_Q2_u_u_M1_L6_aaqaaq_00000k_L4_Wgn_qaqa_L3_R_M1_L3_RE_L3_RE_L4_Wgn_aqaq_L3_R_M1.

and next the additional parts are extracted:

A:A4_M1_L3_RE_00k_M1_L3_RE_00k_M1_L3_RE_0k0_M1_L3_RE_00k,
B:A1_K1,
C:A2_M1_L3_RE_00k_M1_L3_RE_00k.

Aplying the string combining transformation to the string Q4_Q2_u_u_u_u_Q2_u_u_M1_L6_aaqaaq_00000k_L4_Wgn_qaqa_L3_R_M1_L3_RE_L3_RE_L4_Wgn_aqaq_L3_R_M1 and the string that represents the cyclic part, the following strings that represents objects shown in Fig. 7.15 are obtained.

1:"Q4_Q2_u_u_u_u_Q2_u_u_M1_L6_aaqaaq_00000k_L4_Wgn_qaqa_L3_R_M1_L3_RE_L3_RE_L4_Wgn_aqaq_L3_R_M1"
2:"A1_Q4_Q2_u_u_u_u_Q2_u_u_M1_L6_aaqaaq_00000k_L4_Wgn_qaqa_L3_R_M1_L3_RE_L3_RE_L4_Wgn_aqaq_L3_R_M1_M1_L3_RE_0k0"
3:"A1_Q4_Q2_u_u_u_u_Q2_u_u_M1_L6_aaqaaq_00000k_L4_Wgn_qaqa_L3_R_M1_L3_RE_L3_RE_L4_Wgn_aqaq_L3_R_M1_L4_R"
4:"A2_Q4_Q2_u_u_u_u_Q2_u_u_M1_L6_aaqaaq_00000k_L4_Wgn_qaqa_L3_R_M1_L3_RE_L3_RE_L4_Wgn_aqaq_L3_R_M1_M1_L3_RE_00k_M1_L3_RE_00k"
5,6,7:"A2_Q4_Q2_u_u_u_u_Q2_u_u_M1_L6_aaqaaq_00000k_L4_Wgn_qaqa_L3_R_M1_L3_RE_L3_RE_L4_Wgn_aqaq_L3_R_M1_M1_M1"
8,9:"A2_Q4_Q2_u_u_u_u_Q2_u_u_M1_L6_aaqaaq_00000k_L4_Wgn_qaqa_L3_R_M1_L3_RE_L3_RE_L4_Wgn_aqaq_L3_R_M1_M1_L3_RE_00k_M1_L3_RE_0k0";
10:"A3_Q4_Q2_u_u_u_u_Q2_u_u_M1_L6_aaqaaq_00000k_L4_Wgn_qaqa_L3_R_M1_L3_RE_L3_RE_L4_Wgn_aqaq_L3_R_M1_M1_L3_RE_00k_M1_M1_L3_RE_0k0"
11:"A3_Q4_Q2_u_u_u_u_Q2_u_u_M1_L6_aaqaaq_00000k_L4_Wgn_qaqa_L3_R_M1_L3_RE_L3_RE_L4_Wgn_aqaq_L3_R_M1_M1_L3_RE_00k_M1_L3_RE_00k_M1";
12:"A3_Q4_Q2_u_u_u_u_Q2_u_u_M1_L6_aaqaaq_00000k_L4_Wgn_qaqa_L3_R_M1_L3_RE_L3_RE_L4_Wgn_aqaq_L3_R_M1_M1_M1_L3_RE_00k_M1_L3_RE_00k"
13."A4_Q4_Q2_u_u_u_u_Q2_u_u_M1_L6_aaqaaq_00000k_L4_Wgn_qaqa_L3_R_M1_L3_RE_L3_RE_L4_Wgn_aqaq_L3_R_M1_M1_L3_RE_00k_M1_L3_R_0k0_M1_L3_R_00k_M1_L3_RE_0k0"
14:"A4_Q4_Q2_u_u_u_u_Q2_u_u_M1_L6_aaqaaq_00000k_L4_Wgn_qaqa_L3_R_M1_L3_RE_L3_RE_L4_Wgn_aqaq_L3_R_M1_M1_L3_RE_00k_M1_L3_RE_00k_M1_L3_RE_0k0_M1_L3_RE_00k"
15:"A6_Q4_Q2_u_u_u_u_Q2_u_u_M1_L6_aaqaaq_00000k_L4_Wgn_qaqa_L3_R_M1_L3_RE_L3_RE_L4_Wgn_aqaq_L3_R_M1_M1_L3_RE_00k_M1_L3_AE_0k0_M1_L3_AE_00k_M1_L3_RE_0k0_M1_L3_AE_0k0_M1_L3_AE_00k".

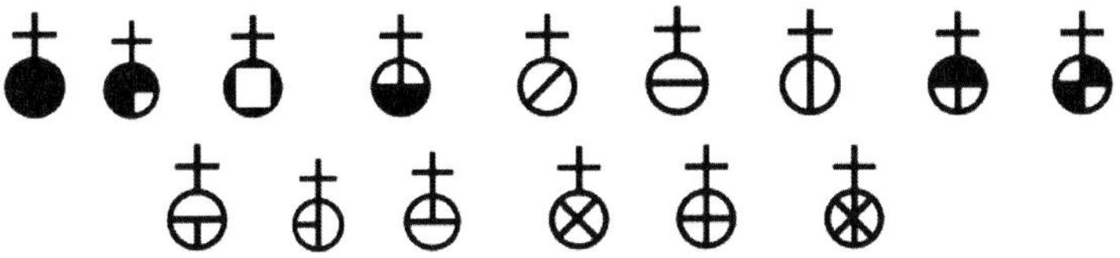

Fig. 7.15. Objects used for testing LP method

For the objects shown in Fig 7.16, the symbolic names were obtained in a similar way.

Fig. 7.16. Objects used for learning LP

1."A1_K1_K1", 2."A2_K1_M1_K1", 3."A1_Q1_u_K1_M1_L3_R",
4."A2_Q1_u_K1_M1_M1_L3_OE_0k0_K1",
5.„A2_K1_M1_L3_O_00k_M1_L3_O_00k " 6."A2_Q1_u_K1_M1_M1_K4",
7."A2_Q2_u_u_K1_M1_L3_RE_L3_RE_M1_L4_Un_qqqa_0k00_L3_RE".

These symbolic names were used to learn symbolic names of objects shown in Fig. 7.17 For example, for the objects shown in Fig. 7.17. a, b, c (first row) the part A1_K1_K1" is combined with

Q4_Q2_u_u_u_u_Q2_u_u_M1_L6_aaqaaq_00000k_L4_Wgn_qaqa_L3_R_M1_L3_RE_L3_RE_L4_Wgn_aqaq_L3_R_M1

and as the result the symbolic name:

"A1_Q4_Q2_u_u_u_u_Q2_u_u_M1_L6_aaqaaq_00000k_L4_Wgn_qaqa_L3_R_M1_L3_RE_L3_RE_L4_Wgn_aqaq_L3_R_M1_K1"

is obtained.

In order to differentiate among objects shown in Fig. 7.17. a,b,c a new specific class is established. This class has two additional attributes "d" and "a". The attribute "d" denotes the difference between the centers of gravity of the hole and a figure. The attribute "a" denotes the angle between the main axis and the line coming through centers of gravity of the hole and the figure.

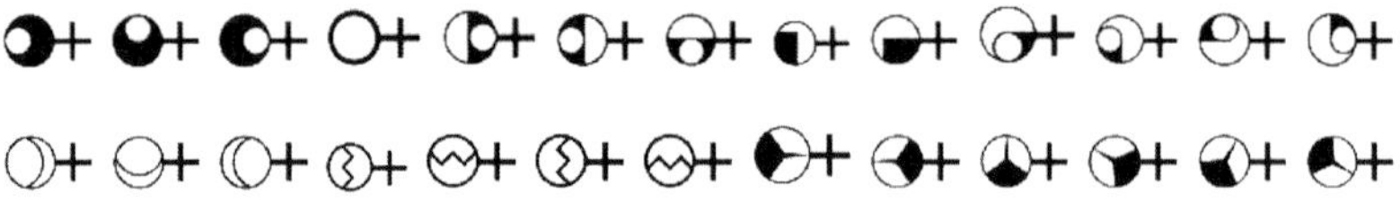

Fig. 7.17. Examples of objects used for learning

7.1.3.5 Learning Parts Decomposition (PD)

Parts decomposition is an important component of the understanding process because all visual objects consist of parts. Often parts of the object cannot be well defined or can be invisible to the naked eyes. The method used for learning objects and their parts is called learning parts decomposition (PD). During learning based on application of the part decomposition method (PD) the most important element of learning is implementation of image transformations used for the decomposition of an object into parts. An implementation of image transformations used for the decomposition of the object into parts is connected with learning of the new perceptual skills that make it possible to understand the new categories of objects. After learning of the new skills the new knowledge is learned. The new knowledge consists of symbolic names of parts that are linked with the symbolic name of the object. For objects that belong to the different categories of the visual objects the different image transformations are implemented. Knowledge of the part decomposition that describes the visual object without reference to the knowledge of the ontological category is stored at the intermediate level of understanding. Parts decomposition method (PD), when object is regarded as the object without reference to the knowledge of the ontological category, can be seen as the reverse to the learning from parts (LP). The symbolic name η that is obtained during LP learning as $\eta(o^i) \equiv \eta(p_1^i) \oplus \eta(p_2^i)$ can be used to learn the symbolic names of its parts $\eta(p_1^i)$ and $\eta(p_2^i)$. However, in order to obtain these symbolic names there is a need to learn the image transformation Θ that makes it possible to decompose the visual object into parts p_1^i and p_2^i.

For example, for an object  knowledge of the part decomposition includes the symbolic names of the parts as well as image transformations that make it possible to divide an object into two parts. During part decomposition the object  is decomposed into part  Q4_u_u_u_u_ L8_aqaqaqaq_L3_R_ L3_R_ L3_R_ L3_R and part  A1_K1_K1. The image transformation, implemented during learning process, is based on the image transformation that divides an object into the circle and the thin part. For the object shown in Fig. 7.18 the image transformation is based on removing of the borders of the convex object. For the object shown in Fig. 7.19 the image transformation is elaborated based on knowledge of the parts of the real world object.

Fig. 7.18. Examples of objects and its parts

Fig. 7.19. Decomposition of the object from the knife category

7.2 Understanding and Learning of the Knowledge of the Sensory Object

Understanding and learning of the knowledge of the visual object was described in Section 7.1. Understanding and learning of the knowledge of the sensory object is described in this Section. As it was described, understanding and learning of the knowledge of the sensory object is considered as acquiring by SUS the complex perceptual skills, whereas the knowledge implementation is concerned with learning of the visual knowledge and non-visual knowledge. The visual knowledge is learned based on the sample of the visual objects that represent the category of visual objects selected for learning. The category of sensory objects, such as the mineral category that is described in this Chapter, is derived from the category of visual objects. A sensory object is an object that is named based on the measurement of values of the different attributes. Visual attributes such as color or shape supply not enough information to obtain reliable recognition and understanding of the sensory object. The category of sensory objects is derived from the category of visual objects or the category of non-visual objects. The category of sensory objects derived from the category of non-visual objects, such as the sound-category, is called the category of non-visual sensory objects. From the sound category the category of speech objects is derived.

The category of sensory objects derived from the category of visual objects, such as the minerals category or the rock category, is called the category of visual sensory objects. The minerals category is the category members of which are objects called minerals. Minerals are objects of study of scientific disciplines called geology. Geology is branch of science that studies of the Earth, including its composition, structure, physical properties, and history. Geology is divided into a number of sub-disciplines. Other disciplines such as mineralogy is concerned with the study of minerals, petrology with the study of rocks, geomorphology with the study of landforms, or paleontology with the study of fossils. The category of minerals and the category of rocks are derived from the category of natural objects (non-man made) and is given by the following categorical chain: $\langle \Pi \rangle \triangleright \langle \sigma_{El} \rangle \triangleright \langle \nu_{\mathrm{Re}O} \rangle \triangleright \langle \nu_{Ear} \rangle \triangleright \langle \nu_{NLiv} \rangle \triangleright \langle \nu_{NMan} \rangle \triangleright \langle \nu_{Min}, \nu_{Rock}, .. \rangle$. Rock is naturally occurring and coherent aggregate of one or more minerals and constitutes the basic unit of which the solid Earth is comprised. Rocks are divided into three major classes according to the processes that resulted in their formation: igneous rocks, sedimentary and metamorphic. Igneous rocks are rocks which have solidified from molten material called magma. Sedimentary rocks are those consisting of fragments derived from preexisting rocks or of materials precipitated from solutions. Metamorphic rocks are rocks which have been derived from either igneous or sedimentary rocks under conditions that caused changes in mineralogical composition, texture, and internal structure. These three classes, in turn, are subdivided into numerous groups and types on the basis of various factors, the most important of which are chemical, mineralogical, and textural attributes. Fig. 7.20 shows examples of members of the minerals category and the rock category. Mineral is any naturally occurring homogeneous solid that has a definite chemical composition and a distinctive internal crystal structure. Although minerals are usually formed by inorganic processes some synthetic equivalents of various minerals, such as emeralds and diamonds, can be produced in the laboratory. Most minerals are chemical compounds and only a small number of minerals (e.g., sulfur, copper, gold) are chemical elements. Minerals are classified into groups based on the identity of its anionic group and the composition of a mineral can be defined by its chemical formula. A mineral is considered to be a crystalline material because it crystallizes in an orderly, three-dimensional geometric form. The crystalline structure of a mineral determines such physical properties as hardness, colour, and cleavage. Minerals are the materials that make up the rocks of the Earth crust. The category of minerals is divided into the native elements category, the sulfides category, the sulfosalts category, the oxides category, the hydroxides category, the halides category, the carbonates category, the nitrates category, the borates category, the sulfates category, the phosphates category, and the silicates category and is given by the categorical chain as follows: $..\triangleright \langle v_{Min} \rangle \triangleright \langle v_{Nel}, v_{Sul}, v_{Sfo}, v_{Oxi}, v_{Hyd}, v_{Hal}, v_{Sfo}, v_{Car}, ... \rangle$. From the minerals category $....\triangleright \langle v_{Min \cdot} \rangle$, the specific minerals categories v_M^i such as malachite, are derived $....\triangleright \langle v_{Min \cdot} \rangle \triangleright \langle v_M^i \cdot \rangle$.

Fig. 7.20. Examples of members of the mineral category (from the Strzelecki collection) and the rock category

The specific minerals categories v_M^i are established and described by the mineralogical scientific body such as the International Mineralogical Association (IMA). Branch of science that studies of minerals is called mineralogy. Mineralogical knowledge, concerned with the mineralogical description can be found in mineralogical books or mineralogical journals. Knowledge of the specific mineral categories is used to elaborate the mineralogical model that is part of the knowledge of the interpretational script of the dictionary mineralogical text. The mineralogical model is described in Chapter 7.3.3.

Understanding of sensory objects from the minerals category requires measurements of many different mineral's properties such as specific gravity, or obtaining a complex data such as data from the X-ray analysis used to compute the values of specific properties such as the cell unit parameters. The physical properties constitute the very useful elements for the recognition and identification of minerals. In some cases examination of the external form (which reflects crystal symmetry) and several physical properties (which depend simultaneously on the chemical composition and on the crystal structure) can lead to proper identification of the mineral. However, usually visual attributes such as color or crystal forms do not supply enough information to proceed with reliable recognition and understanding of objects that belong to the minerals category.

Although a mineral is an object that is a member of the minerals category derived from the category of the visual objects, the visual features such as shape or color can only give reliable information for small part of the minerals categories. During recognition of the mineral, there is a need to use all possible measurements of mineral's attributes. Attributes of the visual object such as size, hardness, luster, color, streak, cleavage, specific gravity, fluorescence, magnetism, radioactivity, tenacity, piezoelectricity, reactivity to dilute acids, reactivity to dilute water have usually the specific meaning in the domain of minerals recognition.

The naming (recognition or classification) of the object from the minerals category, is to assign the examined object o_i to one of the minerals categories v_M^i based on a set of measurements $m(a_i^k)$ and matching process. Matching is finding the mineral category v_M^i for which measures of attributes of the object $m(a^j)$ are matched with values of attributes of the minerals category $u(a_i^j)$. The process of minerals recognition can be written as $o \propto v_M^i : m(a^j) \equiv v(a_i^j)\ , i = 1,\ldots,N$, where N is a number of all specific minerals categories. Mineral's attributes are divided into groups such as the physical attributes a_j^{Ph} or the chemical attributes a_j^{Ch}.

A mineral as the member of the mineral category is given by the name N_i and the pair of characteristic features: the chemical formula and the crystalline structure (φ_i, s_k), and in order to classify an object to one of the minerals categories there is a need to find its characteristic features, chemical composition and crystal structure (φ_i, s_i). The characteristic features (φ_i, s_i) are part of mineralogical knowledge. The value of each mineral attribute a_i^j that is characteristic for each minerals category v_M^i is the result of chemical composition and crystal structure (φ_i, s_i).

During mineral understanding (recognition) there is a need for using special tests and measurements $m(a_i^k)$ that require application of the different measurement methods. Each measurement method specifies factors that are very important ingredients of the measurement process. For each measurement apparatus A, factors such as the type of the minerals sample A^S, knowledge of the apparatus A^K, knowledge of the measurement process A^M, knowledge of the interpretation of the obtained results A^R, are very important ingredients of the measurement process. The measurement of mineralogical attributes can be written as $m(a) \succ A\ \{A^S, A^K, A^M.A^R\}$. Measurement of the specific attributes requires using of a special apparatus and following the special measurement methodology. For example, measurement of the specific gravity of a mineral specimen requires using Jolly balance. Jolly balance is a specific gravity balance that provides numerical values for a small mineral specimen (or fragment) in air as well as in water. Wet tests are qualitative analytical chemical tests, carried out by dissolving mineral in various acids for examination. Wet tests consist in checking to see whether mineral is water soluble, then whether it is soluble in hydrochloric acid, diluted and cold or hot, or concentrated and cold or hot, and finally in nitric acid and in aqua regia.

Flame tests is the qualitative chemical tests which require only very few grains of the mineral to perform analysis. Flame tests provide useful data about the composition of a mineral and narrow the range of identification possibilities considerably. Flame tests consist in heating the mineral in a gas flame (about 1000°C; 1832°F). Flame tests may sometimes be reinforced with oxygen or pressurized air blown from a blowpipe (about 1500°C; 2732°F).

Instrumental methods are chemo-physical methods used for minerals recognition. Instrumental analysis is the best methods used for small samples of minerals. One of the instrumental analysis methods is the electron microprobe. Electron microprobe is a device which can measure extremely small quantities of inorganic material.

Spectroscope is used for qualitative analysis of the elements contained in a mineral on the basis of the presence of absorption lines in the light which has passed through the specimen and then been dispersed by a prism. Infrared (IR) spectrograph measures vibrational frequencies of atomic groups (especially of water or the hydroxyls) present in mineral.

Learning of the knowledge of the category of sensory objects in the context of the development of the system of the mineral understanding "Domeyko" will be described in Chapter 8.

7.3 Understanding and Learning of the Knowledge of the Text Object

The knowledge implementation is concerned with learning of the visual knowledge in the context of the categorical structure of the learned categories of the visual objects and learning of the knowledge that allows the understanding of the content of the text. In this Chapter learning of the knowledge that is utilized during understanding of the content of the text is presented. Understanding and learning of the knowledge of the text object is considered as acquiring the complex interpretational skills of SUS. Acquiring the complex interpretational skills of SUS needed in finding of the meaning of the text is connected with learning of the query-form , the basic-form, the procedural-form and the explanatory and interprapational script.

Learning and understanding are complementary processes. SUS ability to understand depends on the effectiveness of learning process and learning of the new knowledge depends on the SUS ability to understand. Understanding of the text is associated with learning both, knowledge and skills. Understanding of any arbitrary text is a very difficult task. A text can be written in any possible language and it can have very different meanings. One of the important factors of learning is effectiveness of learning. Methods of evaluation of the effectiveness of learning are based on tests that assess the students learning progress. Similarly to a student, SUS will have difficulty in understanding of the text in the area where SUS did not learn skills and knowledge. A student to be able to understand the text needs to learn new concepts and to solve many problems related to the learned topic. Learning needs to be evaluated in order to find if the learning achieved its goal. Evaluation is performed within the framework of the educational assessment. The educational assessment is the process of documenting, usually in measurable terms, knowledge, skills, attitudes and beliefs. The educational assessment is based on measurements of learning outcomes of the individual student in terms of his achievements in a subject domain. The measurement of learning outcomes is based on tests that consist of a number of questions on selected topic.

The very important sources of the knowledge used during learning at school are books and these books are used as the main source of the text-task used during SUS learning. These books consist of problems often called questions, items or tasks. Many books offer solved and supplementary problems that are arranged to allow a progression within each topic. Some books offer the answer for the solution whereas in others every problem has a complete step-by-step explanation for the solution. Complete step-by-step explanation for the solution is very useful in understanding of the problem-solving process. These books usually cover the material that is subject of learning at primary, secondary, or tertiary educational level. As it was described, books contain problems often called questions, items or tasks that are very important source of the text-task used during SUS learning. The category of the text-task is derived from the text category T . The category of the text objects T is the category of visual objects that have dual meaning. The first meaning refers to the visual object such as the book. The second meaning is the meaning of the content of the text. In this book the meaning of the content of the text is regarded as the meaning of object from the category of the text object. The meaning of an object from the category of the text objects is given by the basic-form, the procedural-form, the interpretational and explanatory scripts. The object from the text category is interpreted based on knowledge from the category of science knowledge $\langle \kappa_{KSc} \rangle$ or the category of common sense knowledge $\langle \kappa_{KSK} \rangle$.

In this book, we are focused on understanding of the text that is a short description of a task and consists of a limited number of words. The term 'task' denotes any task (often called problem or question) that can be found in the school handbooks, school tests, IQ tests, university handbooks, or any task such as crossword puzzle questions. For the purpose of this study, a text category T is divided into four different specific categories: the category of text-query T^G, the category of text-task T^T, the category of dictionary-text T^D and the category of long-text T^L. The category of text-tasks T^T is divided into the visual-text-task $T^T[V]$, the action-text-task $T^T[A]$, the explanatory-text-task $T^T[W]$, the educational-text-task $T^T[E]$ and the IQ-text-task $T^T[I]$. In this book only the category of text-query T^G, the category of text-task T^T and the category of dictionary-text T^D are described.

The text-task refers to texts that are used to test the ability or achievements of the students at school. The text-query T^G is the text that does not have the query-part $T^G \equiv T_U^T$, is usually represented by one word or more than one word and is often found as the crossword puzzle questions. The text-query does not have the query part, whereas the text-task has both a task part and a query part, and is denoted as $T^T \equiv T_U^T T_Q^T$. The query-part T_Q^T is a sentence or part of the sentence that has the query term such as "what" or "who". Understanding of the text requires transforming it into the query-form $\Theta(T)$, the basic-form $B(T)$, the procedural-form $P(T)$ and the explanatory script J or interpretational script S. The first task in understanding of the text is to apply the query-form $\Phi(T\ )$ to find the meaning of the text. During understanding each word $w_i \in T$ in the text is transformed into previously learned categories $c_i \in C$ (from the categorexicon) by the categorical transformation $\Re(w_i) \triangleright c_i$. A word $w_i \in T$ in the text can be associated with one or more than one categories $\Re(w_i) \triangleright c_i^1,...,c_i^H$, where H is a number of categories attached to a given word. The text converted into a set of categories is used to find a query-form $\Phi(T^T) \Rightarrow \theta$. The query-form $\Phi(T\ )$ is used to find the basic-form $B(T\ )$, the procedural-form $P(T\ )$, the explanatory script J or the interpretational script S. The query-form $\Phi(T\ )$ is learned during the learning stage.

The text is interpreted in terms of its meaning. Meaning for a dictionary-text is usually given by an interpretational script, whereas meaning of the text-query and text-task is given by both the interpretational or the explanatory script and the basic-form. Understanding of the dictionary-text T^D is based on previously learned interpretational script S, whereas understanding of the text-task T^T is based on the explanatory script J, the basic-form B and the procedural-form P. Understanding of the text-task requires to find the basic-form B and to identify a type of the solution based on the procedural-form P. Understanding of the text-task usually requires also to identify irrelevant parts of the text-task based on an explanatory script J.

The text-query and the text-task are texts meaning of which refers to finding the solution in the form of searching for facts (query-text or geographical text-task), computating of the solution (physical-text-task, mathematical-text-task), seeing the solution (the visual-text-task), undertaking action (action-text-task) or explaining the object (explanatory text-task). The meaning of dictionary text such as biographical dictionary text refers to facts that can be understood in the context of the interpretational script. These texts do not ask for finding of the unique solution. The understanding of the text is often connected with translation of the text from one language to another or asking questions connected with content of the text.

Translation is often regarded as the part of the understanding process. During translation, the source language, i.e. the text to be translated is transformed into an intermediary form, an abstract language-independent representation and next into the final form. The translation into different languages can be performed on any text. In our approach translation consists of two stages: understanding of the text and translation of the text into a selected language. At the first stage, a text is translated into the basic-form (the most general script). For example, translation the text "franklinite is a zinc manganese iron oxide, has a black color and is member of the spinel group" into Polish requires, at first, to transform the text into the basic-form $\varphi_j^N \aleph_j \{c_1^{\aleph},...,c_N^{\aleph}\} val(a(C)_j^{Ph}) \equiv \{c_1^C,...,c_N^C\}(v)$. The meaning of symbols used in description of the basic-form is explained in Section 7.3.3. Next, the basic-form it is translated into an intermediary form "Min is ChemCom has C color and is Mem of G Group". This intermediary form is translated into an editorial form "Min jest chemZwi posiada kolor K i jest Czl G grupy". At the end, the editorial form is translated into a final form "Franklinit jest tlenkiem cynkowo zelazowo magnezowym ma kolor czarny i nalezy do grupy spineli". The final version can be given in many different forms such as: "Franklinit jest koloru czarnego nalezy do grupy spineli jest tlenkiem cynkowo zelazowo magnezowym", or „Franklinit jest tlenkiem cynkowo zelazowo magnezowym nalezacym do grupy spineli i posiada czarny kolor".

7.3.1 Understanding and Learning of the Knowledge of the Text Object – The Text-Query

The text-query category T^G or text-query T^G is the category derived from the text category T. The text-query T^G is the text that does not have the query-part $T^G \equiv T_U^T$, whereas the text-task has both the task part and query-part $T^T = T_U^T T_Q^T$. The text-query $T^G \equiv T_U^T$ is usually represented by one word or more than one word. The text-queries are often the crossword puzzle questions or parts of definitions. It can be shown that the text-query can be easily converted to the text-task by adding the query-part $T^T \equiv T_U^T \oplus T_Q^T$. For example, the text "What does it mean a rectangle is a quadrilateral whose two sides are parallel" can be divided into the query-part "What does it mean a rectangle" and the text part "a rectangle is a quadrilateral whose two sides are parallel". The common part, a rectangle, is the object that is defined.

The text-query that is asking for a synonym or a list of synonyms is called the synonym-text-query $T^G[syn]$. The synonym text-query $T^G[S]$ has usually only one or two words (name) and is often used in crossword puzzles. A synonym is the result of naming process that maps the name from a set of all names $n^k \subset N$ of a given language L_i into the category from a set of all categories $c^k \subset C$.

A synonym can be defined as a set of names (words) that refers to the same category $[n_1^k....n_m^k]->c^k$. The proper synonym is a set of names (words) that refers to the same category $[n_1^k....n_m^k]->c^k$ and each name is given by the one word $\bar{n} \triangleright [w]$. The name $n^k \subset N$ is a word $w_k \in W$ or a group of words $w_k^1...w_k^h \in W$ expressed in the language $l_i \in L$, where N is set of all names, W is a set of all words and L is a set of all languages. A proper name $\hat{n} \in \hat{N} \subset N$ is a name that is given by the one word. A compound name $\bar{n} \in \bar{N} \subset N$ is the name expressed in the form of more than one word $\bar{n} \triangleright [w^1...w^n]$, where n denotes a number of words.

The synonym-text-query $T^G[syn]$ can be transformed into the text-task by adding the query-part such as "what is another name for" $T^T[syn] \equiv T^G[syn] \oplus T_Q^T[syn] = T_U^T T_Q^T[syn]$, where $T^G[syn] \equiv T_U^T[syn]$. For example, the text-query "capsule" can be transformed into the text-task "what is another name for a capsule".

The text-query that refers to the different categorical levels of the categorical chain is called the categorical level text-query $T^G[K]$. The categorical level text-queries $T^G[K]$ usually have a very concise form and are often used in crossword puzzles. For example, the text-query 'a general term for beef, lamb, venison' refers to the general category "meet" from which the specific categories such as ->[beef, lamb, venison, ...] are derived. The categorical chain of the visual objects is given as the hierarchical structure of the categories: $v_o \triangleright \langle v \rangle .... \triangleright \langle v \rangle \triangleright \langle v, v, ..., v, \rangle$, where the ontological categories of the visual object are derived from the visual object category v_o. The categorical chain for the specific categories derived from "meat" is as follows: $... \triangleright \langle v_{meat} \rangle \triangleright \langle v_{bef}, v_{lamb}, v_{ven}, \rangle$.

The text-queries often follow the specific pattern that consists of the marker words such as "type of" or "part of". The text-query that refers to the marker words such as "type of" or "breed of" are called the type text-query $T^G[typ](C_i)$. Similarly, the text-query that refers to the marker words such as "product of" is called the product text-query $T^G[prd]$. The text-query that refers to the marker words such as "ingredient of" is called the ingredient text-query $T^G[ing]$. The text-query that refers to the marker words such as "form of" is called the form text-query $T^G[form]$. The text-query that refers to the marker words such as "part" is called the part text-query $T^G[part]$. The marker words such as "part of" refer to members of any ontological category that consists of the different elements (objects) and are called complex objects. The elements from which the complex object is built are called parts. In the categorical chain the derivation of the part category is denoted as $\langle v_{Tre} \rangle \succ [\tau_{Lef}]$, where symbol $\succ$ denotes that the part category $[\tau\ ]$ is derived from one of the ontological categories $\langle v\ \rangle$. The part category indicates that the derived specific category refers to the object that is part of the complex object. The part category is introduced to represent the different parts of the object. The part category is an auxiliary category that can be derived from any part of the categorical hierarchy. For example, the tree-category consists of the different specific categories given by the following chain: $..\langle v_{Pla} \rangle \triangleright \langle v_{Tre} \rangle \succ [\tau_{Rot}, \tau_{Stm}, \tau_{Lef}, \tau_{Flw}, \tau_{Frt}, \tau_{Sed}]$. Each part category such as the roots category τ_{Rot}, the stems category τ_{Stm}, the leaves category τ_{Lef}, the flowers category τ_{Flw}, the fruits category τ_{Frt}, and the seeds category τ_{Sed}, refer to the parts of the tree. The part can be treated as the independent object that in turn consists of other parts. For example, the tree fruits category is divided into the category of plums, apples or pears: $..\langle v_{Pla} \rangle \triangleright \langle v_{Tre} \rangle \succ [\tau_{Frt}] \triangleright \langle v_{Plu}, v_{App}, v_{Pea}.. \rangle$. A composite object is an object that is assembled from well-defined parts. Objects from which the composite object is assembled are called components. The PartCategory category is related to the ActConsist category.

The text-query, that describes a new category specified by values of one or more attributes of this category, is called the attribute text-query $T^G[C_i](p_1,...,p_N)$. The attribute text-queries $T^G[C_i](p_1,...,p_N)$ are formulated by assigning to the name of the general category C_i the adjective that represents the characteristic value of selected attributes $p_1,...,p_N$. The attribute-text-queries $T^G[C_i](p\ )$ can have one attribute, two attributes $T^G[C_i](p_1, p_2)$, three attributes $T^G[C_i](p_1, p_2, p_3)$ or more than tree attributes $T^G[C_i](p_1,...,p_N)$. The attribute-text-queries $T^G[C_i](p\ )$ refer to the category of visual objects such as "hand tool", "small animal", "small river", "road sign" that are derived based on the specification of the values of one of their attribute $[C_i](p\ )$. The qualitative attribute has its value given as a set of categories $val(p\ ) = \{c^1,...,c^N\}$ and the new category is specified by selection of the one category c^i. Selecting, the ith- category c^i will result in derivation of the new category $[C_i](val(p\ ) = c^i)$. Similarly, in the case of two attributes, selecting for the first attribute the ith- category c_1^i, and for the second attribute the jth- category c_2^i will result in derivation of the new category $[C_i]([val(p_1) = c_1^i, val(p_2) = c_2^j])$.

The text-query that is given in the form of definitions is called the definition-text-query $T^G[Def]$. The definition-text-queries $T^G[Def]$ are usually found in crossword puzzles. The definition-text-query is obtained from the definition by dropping part of the definition which is the name of the defined object. For example, for the definition "nail is small sharpened metal object with broadened flat head" the definition-text-query is obtained by dropping the part "nail is" so the text-query has the form "small sharpened metal object with broadened flat head". Similarly, the text-task can be obtained by adding the query-part "what is the name of" and the text-task obtained has the form "what is the name of the small sharpened metal object with broadened flat head".

Examples of queries in a form of definitions taken from the crossword puzzles are "last prime minister of British Guiana", "tool for lifting loose material", "mandarin orange with a deep reddish-orange peel". Each definition text-query $T^G[Def]$ can be rewritten as the definition of the text-task, for example, "last prime minister of British Guiana"-> "who was the last prime minister of British Guiana". Examples of the definition text-queries $T^G[Def]$, that are given in the form of the definitions, are text-queries in the "tools" domain. The definition text-query $T^G[Def]$ in the tools domain is based on the definitions that can have the different forms. The meaning of the definition text-query $T^G[Def]$ is represented by the basic-form. In order to show the meaning of the definition text-queries, the definitions text-query are presented with its meaning represented by the basic-forms. The simplest definition text-query $T^G[Def]$ is based on the definition type "consists of" $\Im^{Cons} \equiv OthatConsist[P_1 ,..., P_N]$, that means, (o is O that consists of part P1 and ... and part Pn) as in the example: (hammer is a tool that consists of handle and head). The consist-text-query $T^G[\Im^{Cons}]$ has its basic meaning $OthatConsist[P_1 ,..., P_N]$. The consist-text-query $T^G[\Im^{Cons}]$ can be expressed by applying the marker words "that has" or "with" e.g. (hammer is a tool (that {has, consists})(with) handle and head). The query-form of the text-query $T^G[\Im^{Cons}]$ is given as "CTool CConsist CPartTool", where a coding category CConsist includes marker words CConsist={that consists of, that has, with}. The made-text-query $T^G[\Im^{Made}]$ is based on the definition type "made of" $\Im^{Made} \equiv Omade[M]$, that means, (o is O that made of material M). The made text-query $T^G[\Im^{Cons}]$ has its basic meaning $Omade[M]$ and the query-form is given as "CTool CMade CMaterTool". The used-material text-query $T^G[\Im^{Use}(M)]$ is based on the definition type "used for" $\Im^{Use}(M) \equiv Oused[A \rhd M]$ that means (o is O that is used for performing action A on material m). The used text-query $T^G[\Im^{Use}(M)]$ can be expressed by applying the marker words "for", "that is used for". For example, "saw is (tool) for cutting (wood)" or "saw is (tool) used for cutting (wood)" or "saw is (tool) that is used for cutting (wood)". The basic-form is given as $Oused[A \rhd M]$ and the query-form is given as "CTool CUsed CActTool CMaterial", where the coding category CUsed={for, used for, that is used for}. The used-result text-query $T^G\left[\Im^{Use}(R)\right]$ is based on the definition type "used for" $\Im^{Use}(R) \equiv Oused[A \triangleright R]$ that means (o is O that is used for performing action A to obtain result R) e.g. (die is a tool used for cutting thread). The basic-form is given as $Oused[A \triangleright R]$ and the query-form is given as "CTool CUsed CActTool CResult", where the coding category CUsed={for, used for, that is used for}. The used-as text-query $T^G\left[\Im^{UseAs}\right]$ is based on the definition type "used as" $\Im^{UseAs} \equiv OusedAs[T]$ that means (object o is object O used as tools) e.g. (crowbar is a tool used as a lever). The basic-form is given as $OusedAs[T]$ and the query-form is given as "CTool CUsedAs CMTool".

The complex definition can refer to the different aspects of the defined object, for example, "nail used to attach wooden parts into concrete" or "small wide-headed nail often used to affix carpets fabric and other thin materials" are represented in similar way. In order to learn knowledge from a given domain we need to learn the definition that describes different aspects of the defined object. The variation of the description can be learned as the specification of the general form.

The complex definition text-query $T^G[Def]$ can be regarded as a dictionary text that is described in Section 7.3.3. The dictionary text can have more than one definition and is interpreted based on an interpretational script. The interpretational script interprets the meaning of the dictionary text in the context of the model of the object or process. Understanding of the professional work requires understanding of the definition and description of the working process. These definitions refer to the working-process script that is used to interpret of the dictionary text. To illustrate the interpretation of the complex-definition-text-query in the context of the interpretational script, the example of the working-process script is given. The working process script is based on the working process model. In this model the traditional profession is defined by using the characteristic attributes. The definitions are used to represent knowledge about the work category. The work is performed in the special place such as a company, a factory, a mill, a plant, an office, a school, a shop, or a supermarket, and is denoted as K^i. The place where a product is produced is called a factory and is denoted as K^i_W and the special part of the factory, where the given component is produced, is denoted as k^i. The main attribute of the factory is a final product that is produced W^i_{Res} and resources (material) W^i_{Mat} that are used in a production process P^i_W. The production process consists of many sub-processes P^i_W that are responsible for making components w^i_{Res} of the final product W^i_{Res}. The component w^i_{Res} is made by a worker (professional) w^i who is performing one or more actions u^i. The worker w^i wearing a dress w^i_{Dre} is producing a component w^i_{Res} by using material w^i_{Mat} and tools w^i_{Tol}. The worker possesses a knowledge (qualifications) w^i_{Kno} and skills w^i_{Ski}. The basic factory is a factory where there is one worker that produces one category of a product. As an example can be given a tailor or a shoemaker. For the tailor w^{Tai} the place where work is done is the tailor room K_W^{Tai} the product is the wearing such as a jacket or a dress W_{Res}^{Tai}. The tailor uses material w_{Mat}^{Tai} such as fiber and tools w_{Tol}^{Tai} such as scissors or needle, and performes work u^{Tai} such as cutting material, measuring or sewing. The full definition is given in the form $w^{Tai} \equiv [\Pi^i]\,[u^{Tai}]\,[w_{Mat}^{Tai}]\,[w_{Tool}^{Tai}]\,[w_{\mathrm{Res}}^{Tai}](k^{Tai})([w_{Dre}^{Tai}])([w_{Kno}^{Tai}])([w_{Ski}^{Tai}])$ and from this definition the partial definitions of the tailor can be derived. For example, the tailor can be defined as $w^{Tai} \equiv [\Pi^i]\,[u^{Tai}]\,[w_{\mathrm{Res}}^{Tai}]$, that means the tailor w^{Tai} is a man $[\Pi^i]$ who is doing u^{Tai} producing w_{Res}^{Tai}. In similar way a baker can be defined as $w^{Bak} \equiv [\Pi^i]\,[u^{Bak}]\,[w_{\mathrm{Res}}^{Bak}]$, that means the baker [is][person] [who] [bakes] [loaves]. The product of baker work w_{Res}^{Bak} is called loaves and consist of many different products such as bread or rolls $w_{\mathrm{Res}}^{Bak} = \{w_{\mathrm{Res}}^{Bak}(Bre), w_{\mathrm{Res}}^{Bak}(Rol), \ldots\}$ so in definition we can substitute one of the particular products $w^{Bak} \equiv [\Pi^i]\,[u^{Bak}]\,[w_{\mathrm{Res}}^{Bak}(Rol)]$. The baker operation (action, function, activity) u^{Bak} consists of main operations and additional operations. The baker operation is given as $u^{Bak} = \{[u^{Bak}](u_{P1}^{Bak}, \ldots, u_{PN}^{Bak})\}$, where the main operation is called baking and additional operations depend on the final product that is to be obtained. In the case of the given particular product such as A-rolls $w_{\mathrm{Res}}^{Bak}(A\_Rol)$, there is a material $w_{Mat}^{Bak}(A\_Rol)$, tools $w_{Tol}^{Bak}(A\_Rol)$ and knowledge $w_{Kno}^{Bak}(A\_Rol)$ that is connected with making this particular product. Knowledge $w_{\mathrm{Res}}^{Tai}(A\_Rol)$ is a description of the ingredients $w_{Mat}^{Tai}(A\_Rol)$ and operations $u^{Bak}(A\_Rol)$ that need to be applied in order to obtain the final result: $w^{Bak}(A\_Rol)$.

7.3.2 *Understanding and Learning of the Knowledge of the Text Object – The Text-Task*

The text-task category T^T is the category derived from the text category T. The text-task T^T is the text T that can be found in the school textbooks, school tests, IQ tests, university handbooks. The text-task is given in the form of questions, problems or tasks. Usually, meaning of the text-task T^T consists of the two different parts: meaning of the text in terms of the real world situation (phenomena) and meaning in the terms of the task that needs to be solved. The text-task can have different forms, can consist of different categories and can refer to the different phenomena. The text-task can be given in the form of question, query or short description. The text-task T^T is divided into the visual-text-task $T^T[V]$, the action-text-task $T^T[A]$, the explanatory-text-task $T^T[W]$, the educational text-task $T^T[E]$ and the IQ-text-task $T^T[I]$. The category of the educational-text-task is further divided into specific categories based on the type of the knowledge of the subjects learned at school such as mathematics or physics. The mathematical-text-task $T^T[E(Mat)]$ is the educational text-task $T^T[E]$ that can be found in mathematical texts such as handbooks, supplementary books or examination tests. Similarly, the physical-text-task $T^T[E(Phys)]$ is the educational text-task $T^T[E]$ that can be found in physical texts such as handbooks, supplementary books or examination tests. In this book only the category of text-queries T^G, the category of text-task T^T and the category of dictionary-texts T^D are presented. The IQ-text-task $T^T[I]$ is the text-task T^T is that can be found in the IQ test that is used to test of the different abilities such as verbal reasoning (the verbal reasoning tests), numerical ability (the numerical ability test), and diagrammatic or spatial reasoning (the spatial reasoning test).

Meaning of the text-task consists of the two different parts: meaning of the text in terms of the real world situation (phenomena) and meaning of the text in terms of the task that needs to be solved. The first part, that is concerned with the interpretational meaning $\beta^J(T^T) \triangleright J$, is given by description of the stereotypical situation in the form of the script given at the different levels of description that reveals the different levels of details J. The second one, the basic meaning $\beta^B(T^T) \triangleright B$, requires first to transform the text into the basic-form B and then to identify the type of solutions by transforming it into the procedural-form.

The basic meaning of the text-task is given by the meaning of the basic-form and is represented as $\beta^B(T^T) \triangleright B$, where $T^T = T_U^T T_Q^T$, and T_U^T denotes a task part and T_Q^T denotes a query part. The text-task that have the same basic meaning can have a different query part, having the same task part $\beta^B(T_U^T T_{Q1}^T) \equiv \beta^B(T_U^T T_{Q2}^T) \triangleright B$, it can have the different task parts, having the same query parts $\beta^B(T_{U1}^T T_Q^T) \equiv \beta^B(T_{U2}^T T_Q^T) \triangleright B$; or it can have both $\beta^B(T_{U1}^T T_{Q1}^T) \equiv \beta^B(T_{U2}^T T_{Q2}^T) \triangleright B$ a different query-part and task part. The two text-task have the same basic meaning if both refer to the same basic-form $\beta^B(T_{U1}^T T_{Q1}^T) \equiv \beta^B(T_{U2}^T T_{Q2}^T) \triangleright B$. Similarly, the two text-task that differ in query-part $\beta^B(T_U^T T_{Q1}^T) \equiv \beta^B(T_U^T T_{Q2}^T) \triangleright B$ or text part $\beta^B(T_{U1}^T T_Q^T) \equiv \beta^B(T_{U2}^T T_Q^T) \triangleright B$ and refer to the same basic-form B have the same basic meaning $\beta^B(T^T)$. The interpretational meaning of the text-task is given by the explanatory script $\beta^J(T^T) \triangleright J$. The interpretational meaning makes it possible to explain why some parts of the text-task are not relevant in solving of the text-task.

In this book, we are focused on understanding and learning of the educational-text-task. Meaning of the educational-text-task depends on the knowledge domain to which the text belongs. Depending on the knowledge domain to which the educational-text-task belongs, the text-task can be divided into the mathematical-text-task $T^T[E(Mat)]$, the physical-text-task $T^T[E(Phy)]$, the chemical-text-task $T^T[E(Chm)]$, the biological-text-task $T^T[E(Bol)]$ or the geographical-text-task $T^T[E(Geo)]$. The mathematical-text-task $T^T[E(Mat)]$ that does not refer to the real world scene or phenomenon but only to the computation of the mathematical problem is called the mathematical-abstract-text-task $T^T[E(MatA)]$. The basic meaning of the mathematical-abstract-text-task $\beta^B(T^T[E(MatA)]$ is equal to the meaning of the mathematical-text-task and is given by the basic-form $\beta^B(T^T[E(MatA)]) \equiv \beta\ (T^T[E(Mat)]) \triangleright B$.

Understanding of the text-task T^T requires transforming it into the following forms: the query-form $\Theta(T^T)$, the basic-form $B(T^T)$, the procedural-form $P(T^T)$ and the explanatory script $J(T^T)$. Let's consider the following text-task: "Kate bought four apples for $ 2 each; what was total cost" as an example, based on which understanding of the text-task by SUS will be explained. At first SUS transforms this text into the query-form by transforming each word in the text into the previously learned coding categories (from the categorexicon). The words in the text can be associated with one or more than one categories. For example, the word "apple" can be associated with the category "CatBotanyObj", "CatFruit", or "CatMarketObj", that means, it is considered as the object of the botanical science, the fruit, or the object that can be sold on the market. The text transformed into a set of categories is used to learn the query-form. In the case of our example, the query-form is given as CMAN CActBUY C_N COBJ Cfor C_N C_cU Ceach CQ_What C_IS CPrTOTAL CCOST. During understanding process this query-form is used to find the basic-form, procedural-form and explanatory script.

Understanding in terms of the real world situation is interpreted (at the first interpretational level) based on the learned explanatory script "buying_type_0". The first level of the explanatory script "buying_type_0" is given as "Subject bought n objects Each object cost c $ Total cost is C $ Subject need to pay C $ Subject has enough money to pay Subject spend C $ Subject has n objects Subject has less money". This script, although refers to the mathematical task (computation of the cost), does not indicate the way in which the task could be solved. The explanatory script supplies basic knowledge needed to explain meaning of the text in terms of the real world situation. Understanding of the text-task at the first interpretational level refers to knowledge of the scene that it depicts. For example, the task "Kate bought four apples for $ 2 each what was total cost" refers to knowledge of "buying" and describes one aspect of the scenario of 'buying'. The query-form is CMAN CActBUY C_N COBJ Cfor C_N CcU CQ-What C_IS CPrTOTAL CCOST. The explanatory script identifies the categories that are irrelevant to the process of solving task. The text-task is transformed based on the explanatory script "buying_type_0" into form "Kate bought 4 apples Each apple cost 2 $ Total cost is C $ Kate need to pay C $ Kate has enough money to pay Kate spend C $ Kate has 4 apples Kate has less money". Based on this explanatory script the explanatory form is obtained as follows: "Subject: Kate; Object: apple; Action: buy".

The explanatory form is used to identify the irrelevant information and to explain why this information is not needed in the process of finding of the solution.

During the second part of understanding, the query-form is used to find the basic meaning of the text-task (a basic-form). The basic-form is represented as: "Compute cost of n objects if one object costs c $". This basic-form is used to find of the procedural-form that contains the formula to perform the computation of the solution. Interpretation, based on an explanatory script is needed to explain, for example, why some information included in the text is redundant and not useful in solving of the text-task. In the case of the task "Kate bought four apples for $ 2 each what was total cost" the information such as the name of the subject "Kate" or the name of the object that is bought "apple", are redundant. During learning stage (at the intermediate level used to identify the basic-form) these terms are exchanged by letters "A bought four O for $ 2 each what was total cost". In this type of text-task information about "who is buying" is not relevant so the task is transformed into form "four O for $ 2 each what was total cost" and at the end into the form "O cost $ 2 each Compute cost 4 O" where "total cost"="cost of 4 O" and "what was"="Find".

The basic-form "O costs c $ each. Compute cost of n O" or "Compute cost of n objects if one object cost c $" is used to transform this text-task into a procedural-form. The procedural-form is used in the final stage of understanding. During this stage data are extracted, the proper formula is found and the computations are performed. The data: a number of objects n=4, the cost of one item (object) c=2 and the units [$] are extracted and the formula C=c*n for computing the total cost is applied. The procedural-form can be written in the concise form as "Data: c[$], n; Compute cost C: C=c*n [$]. After computation, the results are presented by application of the final-form. The schema of understanding of the text-task shown in Fig. 7.21 illustrates the process of learning and understanding of the text-task.

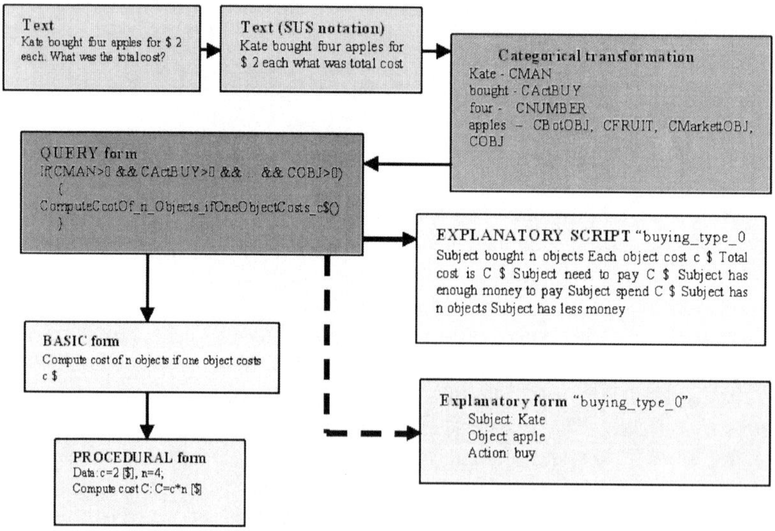

Fig. 7.21. Schema of understanding of the text-task

A text-task is a text meaning of which refers to the unique solution. The solution can be obtained by searching for specific facts (geographical-text-task), computing of the solution (physical-text-task, mathematical-text-task), seeing the solution (the visual-text-task), undertaking action (the action-text-task) or explaining of the object (the explanatory-text-task).

In the following chapters the selected categories of text-task will be briefly described. Understanding of the text-task depends on the factor called the difficulty level. The difficulty level will be briefly described in the next section.

7.3.2.1 Understanding and Learning of the Knowledge of the Text-Task - Difficulty Levels

Solving tasks such as the text-query or the text-task depends on the factor called the difficulty of the task or the difficulty level. The difficulty of the task $\delta[T]$ depends on the difficulty of understanding of the categories $\delta[C]$, the difficulty of understanding of the query-form $\delta[\theta]$, the difficulty of understanding of the basic-form $\delta[B]$, the difficulty of understanding of the interpretational script $\delta[J]$, the difficulty of extracting data and finding formula (data procedural-form) $\delta[P_D]$ and the difficulty in computation and unit conversion (computation procedural-form) $\delta[P_C]$. The total difficulty of solving of the text-task is the sum of the partial components $\delta[T]=\delta[C]+\delta[\theta]+\delta[J]+\delta[B]+\delta[P_D]+\delta[P_C]$ or in the vector form $\delta[T]=a_1\delta^1+...+a_N\delta^N$.

The difficulty of understanding of the categories $\delta[C]$ depends on occurrence of unusual categories in the text-task or occurrence of categories from more advanced knowledge in the text-task. For students, it can be the category that is not yet learned or the category that is unusual. For SUS it can be the category that is not learned. The difficulty of understanding of the query-form $\delta[\theta]$ depends on understanding of the categories and specific structure of the text-task. The difficulty of understanding of the basic-form $\delta[B]$ depends on understanding of the query-form. For a student, it will depend on the ability to think in the categories of the model (abstract thinking) and effectiveness of learning of the knowledge from a given domain. The difficulty of understanding of the procedural-form $\delta[P_C]$ depends on the ability of extracting data and finding the formula, and the ability to perform computation and unit conversion. The difficulty of understanding of the procedural-form usually depends on the learned mathematical knowledge or memorised learned facts.

The difficulty of the text-tasks that have the same basic-form depends only on the difficulty of understanding of the categories $\delta[C]$, the difficulty of understanding of the query-form $\delta[\theta]$ and the difficulty of understanding of the explanatory script $\delta[J]$. In the case when the tasks do not have unusual categories, the $\delta[P_C]$ and $\delta[J]$ parts can be omitted and the difference in difficulties will be given by the difficulty in understanding of the query-form $\delta[T] = \delta[\theta]$. For example, the tasks "car travel 10 km in 30 min what was his speed", "student travels 10.0 km in 30.0 min what was her speed", "student driving car travels 10.0 km in 30.0 min what was his speed", have the same basic-form so the difference between the difficulty of these tasks will depend on the query-form $\Delta\delta[T] \cong \delta[\theta]$.

During assessing students' knowledge in a selected domain e.g. physics, using a test, the most important part of assessment process is to test effectiveness of learning of the knowledge. Applied test should be constructed in such a way that the difficulty of the task should depend only on the essential problem and not on the linguistic difficulty of understanding of the task. That means, the difficulty of the task should depend only on understanding of the basic-form and procedural-form $\delta[T] = \delta[B] + \delta[P]$. The basic meaning given by the basic-form makes it possible to find the essential difference among the text-tasks and to evaluate the essential difficulty level of the task. For example, the following four tasks given in the basic-form T_B1: "Object moves at speed v in time t Find distance s", T_B2: Object moves distance s in time t Find average speed", T_B3: "Object moves at speed v1 in time t1 and next at speed v2 in time t2 Find total distance in sU" and T_B4 "Object moves n laps around circular track in time t diameter of track is D Find average speed", are examples that show the factors that has influence on the essential difficulty level of the text-task. The task T_B2 requires using the same formula as the task T_B1 so they have the same basic meaning. The task T_B3 requires additional knowledge that the distance s can be computed as the sum of two distances s=s1+s2 and knowledge of the changing units. The task T_B4 requires additional knowledge about computation of the distance of the circular track.

7.3.2.2 Understanding and Learning of the Knowledge of the Text-Task – The Explanatory Text-Task

The explanatory text-task is the text-task which requires finding the solution in the form of explanation of the phenomenon or object. There are many events, objects and facts which require explanation. Explanation is associated with understanding that is represented by different media such as music, text and graphics. Explanation can be defined as a set of statements, constructed to describe a set of facts, which clarify the causes, context, and consequences of those facts. Explanation can be given in a simple form by indicating the similar objects or phenomena, or in the complex form showing step by step the solution. Explanation can be also given by visualizing an object, a problem or a solution.

The answer to the explanatory text-task can not be easy to evaluate. Usually the answer to the explanatory text-task is based on the interpretational script and in many cases, understanding of the explanatory text-task can be considered as understanding of the dictionary text. The explanatory text-task has its query-part indicating that solution requires giving the explanation such as "What is meant by the term", "Explain relationship between", "Explain why". The explanatory text-task often has basic meaning as follows: "Why does A cause P", "What is meant by the term W", "Explain relationship between A and B", "Explain why A is p".

Here are few examples of the biological text-task that are explanatory text-task: "What is meant by the term cloning?", "Why are mammals more susceptible to a build-up of carbon dioxide than fish?" "Why does hyperventilating cause person to become unconscious?" or "Explain relationship between an allergen, mast cells and histamines".

Some explanatory text-tasks refer to diagrams, for example, "the following diagram outlines events associated with the production of a polypeptide chain in a eukaryotic cell. a. What is the name of the process at step 1?" and some to the figure "Based on Figure 2, suggest how tryptophan prevents repressor protein function". Explanation often refers to a visual object. For example, the crystal habit can be explained by referring to its visual aspect represented by the crystal form. The crystal is a geometric body so it can be defined from a purely descriptive geometric point of view. A characteristic form of a mineral (simple crystalline form) is defined in terms of geometrical solids such as a cubic, a dodecahedral, a rhombohedral or an octahedral. The geometrical solids can be expressed in the form of the symbolic names and used in process of visual understanding to explain the crystalline structure of crystal habit. Examples of the simple crystalline forms, represented in the forms of the symbolic names, are shown in Fig. 7.22 (1 to 8).

Explanations are often given in terms of the similarity of appearance of the objects such as needle-like, tree-like, kidney-shaped, like-thread, resembling bunch of grapes, sheet forms, lens shapes, Cog-Wheel shapes, or honeycomb aggregates.

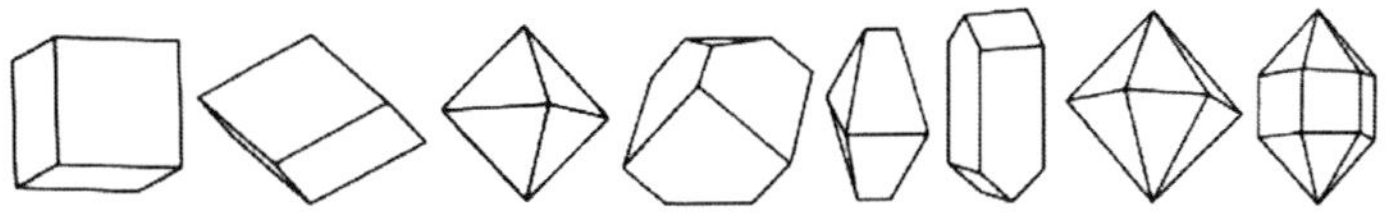

1 2 3 4 5 6 7 8

Fig. 7.22. Examples of the simple crystalline forms, represented in the forms of the symbolic names 1. $\rho[L_N^6]\{L_R^4, 2L_W^4\}$ 2. $\rho[L_W^4]\{3L_W^4\}$ 3. $\rho[L_W^4]\{4L^3\}$ 4. $\rho[L_H^8]\{L_M^5, 2L_G^5, L^3\}$ 5. $\rho[L_Z^6]\{2L_T^4, 2L^3\}$ 6. $\rho[L_X^8]\{2L_R^5, 2L_W^4\}$ 7. $\rho[L_W^4]\{6L^3\}$ 8. $\rho[L_Y^6]\{3L_W^4, 6L^3\}$

Also, explanation can be given in the form of steps that need to be followed to obtain a solution. The following example, from solving a simple equation, illustrates this type of explanation: "substitute the values for the variables into the expression 4(3)+(-8); orders of operation tells you to multiply first [4(3)=12]; Substitute (12)+(-8). Signs different? Subtract the value of the numbers 12-8=4; give the result the sign of the larger value (No sign means +) +4. The value of the expression is as follows 4a+z=4".

Explanation can be given in the form of an animation that explains the abstract concept such as cumulative distribution or method such as regression analysis. Fig. 7.23 shows examples of the animation used to explain statistical concept that is used in the context of SUS explanatory process. Some of the issues connected with visualization in the context of system of integrated packages were presented in [110].

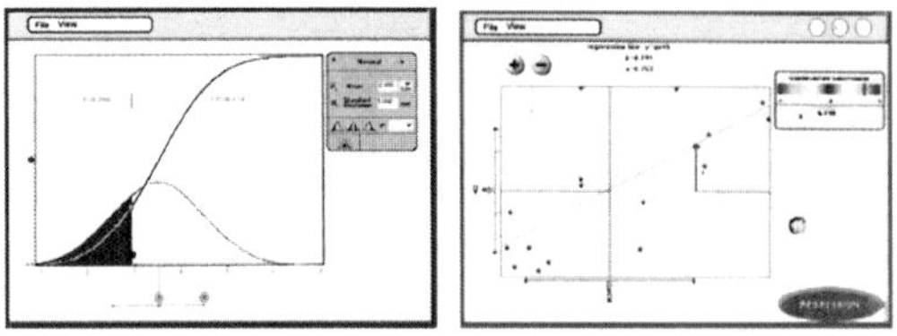

Fig. 7.23. Examples of the concept visualization a. cumulative distribution b. regression analysis

7.3.2.3 Understanding and Learning of the Knowledge of the Text-Task – The Action-Text-Task

An action-text-task is a text-task meaning of which refers to performing an action. An action is prescribed by giving simple commands or steps that need to be followed in order to perform required task. Learning of performing action is to learn steps (commands) that lead to obtain the requested goal. The simplest action-text-task given in the form of the command that needs to be performed is called the command-text-task. For example, 'draw a convex shape' or 'write a time in words' are command-text-tasks given be the command "draw" or "write". The command-text-task such as "draw circle" can be interpreted in the context of the robot that is performing specific task. The robot can use different tools to draw a circle or it can draw it on the screen. To understand the meaning of the action-text-task is to perform a proper action. During learning of the knowledge of the action-text-task, the query-form, the basic-form and the procedural-forms of the action-text-task are learned. For example, a simple action-text-task "write number 45" is learned by designing a procedural-form that describes the procedure write_text(). The Action-Expert implements both knowledge and skills needed to perform the action given by the action-text-task. Learning of the meaning of the action-text-task such as "Draw triangle inside circle', "Draw circle in the box", "Draw star in centre box", "Draw circle in box on right" requires learning of the placement categories such as "in centre of the box", or "inside circle". The procedural-form of these action-text-task is much complex and requires implementation of the special procedure such as the procedure Draw_Fig1_Inside_Fig2_Condition.

The command-text-task is the action-text-task that has the command part. In order to perform a command, the command-text-task needs to be understood. Understanding of the command-text-task, similarly like other text-tasks is based on the previously learned basic-form and procedural-form. The command-text-task can be given by the command part represented by words "circle", "write", "draw", "label", "tick", "measure", "trace", "colour", "link", "join", "divide". Examples of the action-text-tasks used during SUS learning of the command-text-tasks are as follows: "Circle groups of animals 2. Circle the lighter animal. Circle the longer object. Circle the container with the largest capacity. Circle the objects which will roll. Circle the shapes with 4 straight sides. Circle the shape of the cross section. Circle the closed shapes. Circle the shapes which show a line of symmetry. Circle the coins needed to buy an apple for 50c. Label the coins.

The recipes-text-task is the action-text-task that describes all steps needed to achieve the goal, such as recipes. The recipes-text-task is learned in similar way to other text-tasks by learning the query, the basic and the procedural-form. However, the procedural-form is much complex than for any other text-task because it requires to learn all steps needed to teach the robot to perform actions that will achieve the goal. Examples of recipes-text are: "Delicious spaghetti bolognaise Ingredients: ½ packet spaghetti, 2 tablespoons oil, 1 kg minced meat, 2 finely chopped onions, ½ bottle tomato sauce, 250 g bottle tomato paste, salt and pepper to taste. Method: 1. cook the spaghetti in boiling water 2. heat oil in another pot 3. Add meat and chopped onion and cook for twenty minutes 4. Stir in the tomato paste, sauce, salt and pepper 5. Leave to simmer 6. Strain the spaghetti and serve with the sauce".

7.3.2.4 Understanding and Learning of the Knowledge of the Text-Task – The Visual-Text-Task

A visual-text-task is a text-task that has a part that refers to the visual object. The simplest visual-text-task is given in the form of the linguistic description that refers to the pictorial representation. For example, the visual-text-task "Name the shapes" refers to the visual object shown in Fig. 7.24.(a, b, c). The results of understanding of this text are names of the objects shown in Fig. 7.24. (a, b, c). Understanding of this visual-text-task requires finding the name of perceived objects during the naming process. Names are found by invoking the Visual-Procedural-Expert which uses the Naming-Expert to obtain the name of the visual object.

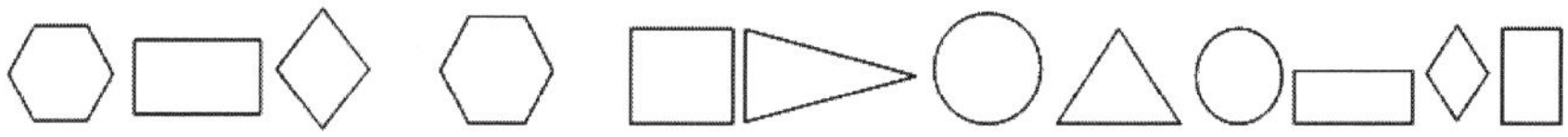

Fig. 7.24. Examples of visual-text-tasks: 1. "Name the shapes" 2. "This is hexagon. How many sides? How many corners?" 3. Colour the rectangles yellow and the circles green

The visual-text-task such as "This is hexagon. How many sides? How many corners?" can be solved in two ways. The first way employs the Visual-Procedural-Expert that invokes the Geometric-Procedural-Expert. The Geometric-Procedural-Expert formulates an answer based on the geometrical knowledge. The geometrical knowledge is represented in the form of the definitions, such as "hexagon is a polygon that has 6 corners and 6 sides". It is important to emphasize, that in this approach, obtaining the answer does not require "looking" at the object. In the second way, the Visual-Procedural-Expert invokes the Naming-Expert and, after examination of the object shown in Fig. 7.24(d), the name "hexagon' is assigned to that object, and next the Geometric-Procedural-Expert is invoked to formulate an answer based on the geometrical knowledge of the named object. This approach can be used to understand the visual text-task "How many corners has object shown?". In this context, it should be noted that understanding of the text-task means to find the solution for this text-task.

Another visual-text-task such as "Colour the rectangles yellow and the circles green" is solved by the Visual-Procedural-Expert that obtains the names of the objects by invoking the Naming-Expert.

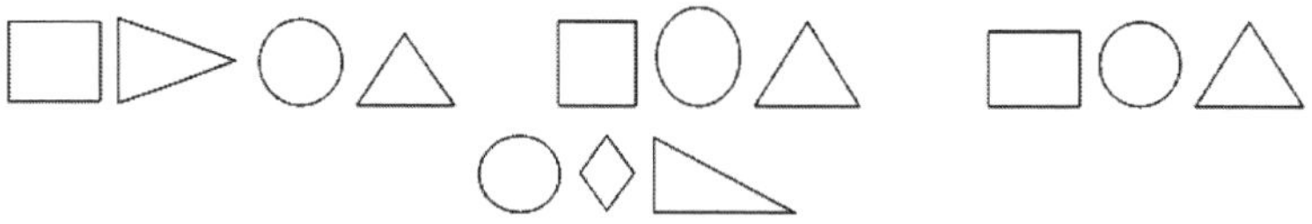

Fig. 7.25. Examples of visual-text-tasks: 1. "Circle the shapes with 4 straight sides" 2. "Circle the shape that has no straight sides" 3. "Colour the shape with no corners" 4. "Colour the symmetrical shapes (Circle the shapes that show a line of symmetry)"

Another group of visual-tasks, shown in Fig 7.25, such as "Circle the shapes with 4 straight sides" "Circle the shape that has no straight sides" "Colour the shape with no corners" can be solved by SUS by "looking" for the special features of the visual object such as corners or straight lines. These tasks can be solved by invoking the Visual-Procedural-Expert and next the Naming-Expert to obtain the names of the visual objects. The named objects are examined by the Geometric-Procedural-Expert and objects such as triangles or rectangles that fulfill conditions "object that has corners" or "object that has straight sides" are selected. In similar way, SUS can solve the following visual tasks shown in Fig. 7.26: "Colour the curved shapes blue" or "Circle the closed shape or Colour the symmetrical shapes".

Fig. 7.26. Examples of visual-text-task 1."Colour the curved shapes blue" 2. "Circle the closed shape"

The visual tasks concerning 3-dimensional objects (3D), shown in Fig. 7.27, such as "How many corners?", "How many surfaces?", "How many edges?", or "What shape is each face – square, triangle, circle?" is solved in similar way to the task concerning 2-dimensional object, by "looking" for the special features of the visual object such as corners, straight lines, or faces.

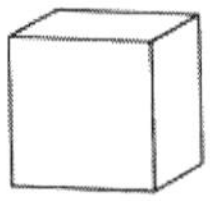

Fig. 7.27. Examples of visual-text-task 1. "How many corners? How many surfaces? How many edges?" 2. "What shape is each face – square, triangle, circle"

The visual tasks, shown in Fig. 7.28, such as: "Draw the circle in the box on the right", "Draw a star in the centre box." are tasks that require understanding of the placement categories such as "in centre of the box", or "inside circle". These tasks can be also regarded as the action-text-task. These tasks are solved by invoking the Visual-Procedural-Expert and next the Visual-Procedural-Expert to find "box on the right" or "centre box". The Action-Expert is invoking the Drawing-Expert to draw circle or star inside selected box.

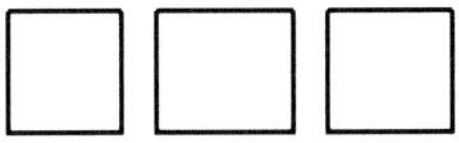

Fig. 7.28. Examples of visual-text-tasks: 1. "Draw the circle in the box on the right" 2. "Draw a star in the centre box"

The visual tasks, shown in Fig. 7.29, such as: "Draw the side view", "Draw from the side", "Draw this model from the top", "Draw the shape you would see from the top", or "Draw the hand from the top view" are solved by the Visual-Procedural-Expert by naming the object and next using knowledge of the Geometric-Procedural-Expert and Drawing-Expert to draw the view of object. The visual knowledge concerning the different views of a given 3-D object is learned and stored as the part of the visual geometrical knowledge of the Geometric-Procedural-Expert.

Fig. 7.29. Examples of visual-text-tasks 1. "Draw from the side" 2. "Draw this model from the top"

The visual-text-tasks are tasks that are often used in intelligence tests. Intelligence tests are series of tasks designed to measure the capacity to make abstractions, to learn, and to deal with novel situations. General intelligence that is measured by tests of 'general intelligence' known as 'IQ tests' has typically meant to measure the ability to see relations in, make generalization from, and relate and organize ideas represented in a symbolic form. Intelligence tests that include tasks that deal with visual forms (shapes) are called the visual intelligence tests. They are divided into several groups: the visual discrimination tests, the visual memory tests, the visual-spatial relationship tests, the visual form constancy tests, the visual sequential memory tests, the visual figure ground tests or the visual closure tests [111], [112], [113] [114], [115], [116], [117]. The intelligence test is designed to measure the different types of mental ability such as verbal, mathematical, spatial and reasoning skills. One of the IQ tests that was developed and has been used in many countries is a progressive matrices test. The progressive matrices test can be described by verbal description: "Look at these figures carefully. Which one from here should go where the empty one is?" The progressive matrices tests have been used to test an ability to find relationships among the visual objects. In these tests two types of relationships are established: the first in the horizontal and the second one in the vertical direction. Fig. 7.30 shows different categories of visual tests used to test general mental ability. The visual tests used for learning of SUS to solve visual tests are described in [1].

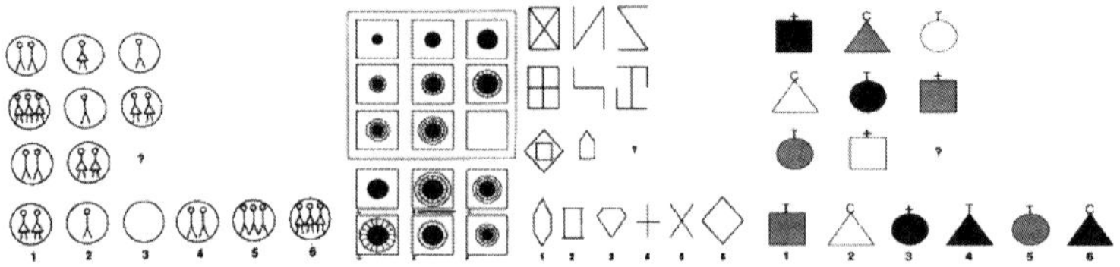

Fig. 7.30. Examples of the visual-text-task used in the intelligence tests

7.3.2.5 Understanding and Learning of the Knowledge of the Text-Task – The Coding Category

Knowledge implementation is concerned with learning of the knowledge that is connected with understanding of the content of the text. In this section coding categories that are important part of the learned knowledge are described. The categories are the most important part of the knowledge structure of SUS. As it was described, there are three types of general categories: the ontological categories of the visual object v_i, the knowledge categories κ_i and the coding categories c_i.

The ontological categories of the visual object v_i, are described in Chapter 6. These categories refer to the visual objects and are part of the visual knowledge represented by the categorical chain. The knowledge categories κ_i are categories of the scientific disciplines such as mathematics, physics or chemistry. Each scientific discipline has its own terms or concepts that are defined within the domain of these scientific disciplines.

The knowledge categories κ_i and the ontological categories of the visual object v_i are established based on existing knowledge. The coding categories c_i are related to both the knowledge categories κ_i and ontological categories of the visual object v_i, and are established during learning of the query-form $\Theta(T)$. The coding category is the category that is established during iterative learning process and usually it is derived from the ontological categories of the visual object v_i or knowledge categories κ_i. These categories are established in order to obtain the generalization of the learned text T. At the first stage of learning, each word w_i in the learned text T is associated with the one or more than one coding categories $\Re(w_i) \triangleright c_i^1,...,c_i^H$, where H is a number of categories attached to a given word. These categories c_i^j as well as association of a category with a given word

$\Re(w_i) \triangleright c_i^1,...,c_i^H$ can be changed during an iterative learning process. The learned categories c_i^j are part of the categorexicon and are stored in the categorexicon file $c_i^j \in C$. The coding categories are divided into the category of visual object $c_i[V]$, the category of action $c_i[A]$, and the category of attributes $c_i[F]$.

The ontological categories derived from the category of visual object are rather well established and are represented by the categorical chain. The coding categories are used to define the query-form. During the coding stage the ontological category needs to be selected at the appropriate categorical level. For example, the categories derived from the figure category such as a polygon category or rectangle category, given by the following categorical chain: $O \triangleright \langle v_{Fig} \rangle \triangleright \langle v_{2DF} \rangle \triangleright \langle v_{Pol} \rangle \triangleright \langle v_{Qud} \rangle \triangleright \langle v_{Rec} \rangle$ are stored in the categorexicon file as follows:

figure	Cat2DFig	CatFigure	
polygon	CatPolygon	Cat2DFig	CatFigure
quadrilateral	CatQuad	CatPolygon	Cat2DFig
	CatFigure		
rectangle	CatRect	CatQuad	CatPolygon
	Cat2DFig	CatFigure	

During learning of the query-form the coding categories can be selected at any categorical level. The word “rectangle” can be coded as Crect, CQuad, CPoly, C2DFig. During learning process the names of the visual categories such as CatRect, CatPolygon, CatQuad, Cat2DFig, CatFig are often learned as names of the coding categories such as CRect, CPoly, CQuad, C2DFig, CFig, by changing part “Cat” into “C’. This way of coding makes it possible to find the optimal level of the query-form that can cover the broad range of the meaning of the text and at the same time do not lead to confusion by mixing the text having the different meaning.

The category of visual object can be associated with one or more than one coding categories. For example, the word “apple” that is the member of the category of the visual object can be associated with the coding categories “CFruit”, “CEatObj”, or “CMarketObj” that means it is considered as the object of the botanical science, the ingredient of the food or an object that can be sold on the market:

apple	CFruit	CBotObj	CBiolObj
apple	CFruitFood	CBotFood	CFood
apple	CFruitMarket	CMarketOBJ	CMarket

In the case when the text belongs to one of the scientific disciplines such as physics the coding categories are part of the categorexicon of the physical domain e.g. categorixicon of physics called PhisCategorixicon. During learning of the query-form for the physical text-task, the word “apple” is associated with the coding categories “CMovObj” that is the category of the physical domain.

7.3.2.6 Understanding and Learning of the Knowledge of the Text-Task – The Query-Form

Learning of the knowledge of the text-object is to learn the coding categories, the query-form, the basic-form, the procedural-form, the explanatory script and interpretational script. In previous Section the coding categories that are used to obtain the generalization of the learned text and to learn the query-form were described. In this section the query-form is presented.

The query-form $\Theta(T^T)$ is the generalization of the text-task expressed in the form of the clauses consisting of the coding categories. As it was described in the previous section, the coding categories are established based on the knowledge categories of the specific scientific domain such as physics, geography or mathematics. The learned query-forms $\Theta(T^T)$ has a form of the clause $if((k_i^1 \in Z_k) \wedge (k_i^2 \in Z_k),...,\wedge(k_i^N \in Z_k))thanB_k$, where B_k denotes the k-th basic-form. The k-th set of categories $Z_k = \{c_i^1,...,c_i^N\}$ that represent the k-th text-task is obtained during learning of the text-task T^T. During learning of the text-task T^T each word of the text $w_i \in T^T$ is coded as the coding category and stored in the categorexicon $c_i \in C$. For each group of text-tasks $\{T_h^T\}$ (h=1, to H) the generalization of these tasks is obtained as a query-form represented by a set of coding categories $Z_k = \{c_i^1,...,c_i^N\}$. A set of coding categories Z_k represent the group of text-tasks $Z_k \propto \{T_h^T\}$.

During the understanding text-task T_o^T, each word of the text-task $w_i \in T_o^T$ is transformed into a set of coding categories $K_o = \{k^1,...,k^M\}$. The understanding is to find the basic-form. This process is represented by reasoning algorithm: For k=1 to K, $if((k^1 \in Z_k) \wedge (k^2 \in Z_k),...,\wedge(k^M \in Z_k))thanB_k$, where K is a number of learned tasks, and B_k denotes the k-th basic-form.

At the first stage of learning, the query-form is given in the form that does not take into account the grammatical dependence of the words. During understanding process, an examined text-task is transformed into the stream of coding categories and classified into one of the basic-forms based on the learned query-forms. For example, the text-task Z1 "car travel 10 km in 30 min what was his speed", Z2 "student travels 10.0 km in 30.0 min what was her speed" are at first transformed into the basic-form B1: "Object O move distance s [s] in time t [t]. Compute velocity v [v]", by application of the following query-forms Z1, Z2: "CObjMov CActMove CNum [CUv] CAuxIn CNum [CUt] CQCalcul CatPhisObjSpd". This query-form does not take into account of the grammatical dependence among the words. The task that is given in the non-grammatical form as in the task Z1a, "10 car speed travel km in 30 his min what was" can be also translated into the query-form B1. At the first stage of learning, it is assumed that the task is taken from one of the books that is used for learning of the knowledge from a given domain e.g. physics. The small alternation of the task, such as in the task Zb "travel car 10 km in 30 min what his speed was" does not lead to error in understanding. This way of understanding of the text is very similar to the way in which our human understanding can cope with these tasks. In the second stage of learning, the frazes are learned in order to keep the grammatical order and in the third stage, the grammatical relations are learned. Learning at the third stage will prevent from transforming tasks such as Z1a into the basic-form B1.

Understanding of the huge varieties of text-tasks that have equal or nearly equal meaning depends on the results of the learning stage. Learning usually does not cover all cases that are needed to interpret precisely the meaning of the text or to avoid misinterpretation. One way to avoid misinterpretation is to learn as many cases as possible to ensure that misinterpretation will not occur. This way of learning is similar to our human learning, where the deep knowledge and experience can guarantee proper understanding. However, when learning is not extensive, especially at the beginning of learning of the knowledge from a selected domain, understanding some of the tasks can be based on utilization of the key categories and making the appropriate assumptions. In some cases it can be rather guessing of the proper meaning than "rational" interpretation.

Learning the query-form is the same for the text-task, the query-task or the dictionary-text. In the case of the text-task that consists of the text part T_U^T and the query-part T_Q^T these parts can be learned independently. The query-form $\Theta(T^T)$ for a given text-task T^T can be expressed by using the different categories or categories at the different categorical level. The different categories can modify the description of the task T_U^T but should refer to the same basic-form B (to have the same basic meaning $\beta^B(T^T)$). For example, the task "Kate bought four apples for $ 2 each what was total cost" can be transformed into the task "STUDENT bought four FRUIT for $ 2 each what was total cost" where the category "Kate" is changed into the more general category "student" ("Kate"-> "STUDENT") and the category "apple" is changed into the more general category "fruit" ("apple"->"FRUIT"). We will write changing categories in the form (c1->c2) that means the category c1 is changed into c2, the category of the higher level. Similarly we can exchange the category ("buy"->"purchase"), ("four"->"4") and ("$"->"dollar"), so we obtain text "STUDENT purchase 4 FRUIT for 2 dollar each what was total cost". By exchanging categories (STUDENT->MAN), (FRUIT->OBJ), (purchase->BUY), (4->N), (dollar->U) we obtain the general query-form that can cover meaning of many text-tasks. The general query-form is "MAN BUY N OBJ for U N What was total cost", where N is the general numeric category and U denotes the general unit category. In this example, the category U can be exchanged for more specific monetary category UP. The task "MAN BUY N OBJ for U$ N What was total cost" covers many text-tasks such as "Student bought four pens for $ 2 each what was total cost", "Student bought 4 pens for 2 dollar each what was total cost", "Student purchased four pens for $ 2 each what is total cost". Exchanging categories is the task relatively easy; however coping with the structural changes of the text-task is rather complex problem. These issues will be discussed in more details in Chapter8.

The text-task T^T consists of the text part T_U^T and the query-part T_Q^T. The query-form $\Theta(T^T)$ is also divided into the text part $\Theta(T_U^T)$ and the query-part $\Theta(T_Q^T)$ so the query-part can be learned independently. The query-part of the query-form $\Theta(T_Q^T)$ that is learned can be added to many different text parts $\Theta(T_U^T)$. For the task, "Kate bought four apples for $ 12 each what was total cost ", the query-part is "what is total cost" and can be exchanged into the form "compute cost", "compute total cost", "compute cost of four pens", "compute cost of all pens", and "compute cost of all bought objects". Learning of the query-form can be learned independently for the text part T_U^T and the query-part T_Q^T. The word or a group of words in the query-part such as "find", "calculate", "what is", or "compute" can be coded by using coding categories such as CQ_CALCULATE.

7.3.2.7 Understanding and Learning of the Knowledge of the Text-Task – The Basic-Form

As it was described, learning of the knowledge of the text-object is to learn the coding categories, the query-form, the basic-form, the procedural-form, the explanatory script and interpretational script. In the previous sections the coding categories and the query-form were described. In this section the basic-form is presented.

The basic-form $B(T^T)$ is the representation of the basic meaning of the text-task. The basic-form is used to interpret the text-task as the problem that needs to be solved. The basic-form is learned and used to identify the problem and can be understood as the computational model of the task. During understanding of text-task T_O^T, each word of the text-task $w_i \in T_O^T$ is transformed into a set of coding categories $K_O = \{k^1,...,k^M\}$ and the basic-form is found by applying the reasoning algorithm: For k=1 to K, $if((k^1 \in Z_k) \wedge (k^2 \in Z_k),...,\wedge(k^M \in Z_k))thanB_k$, where K is a number of learned tasks, and B_k denotes the k-th basic-form. This algorithm was described in the previous section.

The basic-form refers to the basic meaning of the task. The meaning for the text-query and the text-task is given by both the script and the basic-form $\beta(T^G) \triangleright (J,B)$ and $\beta(T^T) \triangleright (J,B)$, and the basic-form contains only the information needed to find the solution. Two text-tasks have the same basic meaning if both refer to the same basic-form $\beta^B(T_1^T) \equiv \beta^B(T_2^T) \triangleright B$. For example, two text-tasks T1: "ship steams at constant velocity of 30 km/h how far does it travel in two days" and T2: "toy train moves along winding track at average speed of 0.25 m/s how far will it travel in 4 minutes " has the same basic meaning B: "Object move at average speed v Find distance after time t".

As it was described in previous sections, the difficulty of the task $\delta[T]$ depends on the difficulty of understanding of the categories $\delta[C]$, the difficulty of understanding of the query-form $\delta[\theta]$, the difficulty of understanding of the basic-form $\delta[B]$, the difficulty of understanding of the interpretational form $\delta[J]$, the difficulty of extracting data and finding formula (data procedural-form) $\delta[P_D]$ and the difficulty in computation and unit conversion (computation procedural-form) $\delta[P_C]$. However, the difficulty of the text-task that have the same basic-form depends only on the difficulty of understanding of the categories $\delta[C]$, and the difficulty of understanding of the query-form $\delta[\theta]$. The basic meaning given by the basic-form makes it possible to find the essential difference among text-tasks and to evaluate the difficulty level of the task. For example, basic-forms B2: "Object move at speed v in time t Find distance s", B3: "Object move at average speed v Find distance after time t", B4: Object move distance s in time t Find average speed", B5: "Object move at speed v1 in time t1 and next at speed v1 in time t1 Find total distance in sU" and B6 "Object move n laps around circular track in time t diameter of track is D Find average speed" are examples of basic-forms that show factors that influence the difficulty level of the text-task. The difficulty level of the tasks T2 and T3 is the same. The task T4 requires to transform the formula used in solving tasks T2 and T3. The task T5 requires additional knowledge that distance s can be computed as the sum of the two distances s=s1+s2 and the knowledge how to change units. The task T6 requires additional knowledge about computation of the distance of the circular track.

For the text-query the basic-form is a very similar to the query-form. During understanding process the query-form is used to find the basic-form. For example, for the text-query $T^G[typ](C)$ or $T^G[part](C)$ the basic-form is typ_o (object type) or $part\_o$ (object parts) such as a car type or part of camera. The text-queries from the geographical knowledge domain such as "river in Poland", "mountain in Hungary", "lake in Armenia", "island in France" are transformed into the query-form GeOBJinDOMAIN and into the basic-form o_in_X (object o in the container X). Similarly, the text-query such as "country with highest (lowest) population in Europe" is transformed into the basic-form $o\_with\_Max(a)\_in\_X$ (object with max value of attribute a in container X). The text-queries such as "population of Poland" or "Capitol of Poland" are transformed into the basic-form $Q(a)\_in\_X$ (value of attribute a in container X). The text-query such as "country with high population" is transformed into the basic-form $o\_with\_H(a)$ (object with high value of attribute a).

The basic-form for the text-task consists of two parts: the text part and the query part. The basic-form can be differently formulated however it will always follows the simple scheme as this: "Compute cost of n objects if one object costs c $". The basic-form is strictly connected with procedural-form into which the text-task is transformed during the stage of finding of the solution. In order to find the solution, in most cases the transformation into the procedural-form is sufficient to understand the task and find the solution. The procedural-form usually, supplies the computational model of the task and the knowledge how to compute the solution. For each scientific domain such as mathematics or physics the computational model is expressed in the form of well-defined concepts. For example, the basic-forms of the mathematical-text-task "O cost c $ each Compute cost of n O" or "Compute cost of n objects if one object cost c $" are used to translate the text-task into the procedural-form.

7.3.2.8 Understanding and Learning of the Knowledge of the Text-Task – The Procedural-Form

In the previous sections the coding categories, the query-forms and the basic-forms were described. These forms were learned to make it possible to understand the text-task and to solve the problem represented by the text-task. In this section, the procedural-form which plays an important role in finding a solution to the text-task is described.

The procedural-form $P(T^T)$ is the description of steps that lead to the solution of a text-task when using the mathematical models or searching procedures. It is used at the final stage of understanding of the text-task. The procedural-form $P(T^T)$ that represents a solution that is based on a mathematical model is called the computational-procedural-form. The procedural-form $P(T^T)$ that represents the solution that is based on a searching procedure is called the searching-procedural-form.

For the text-queries with searching-procedural-form, learning of the procedural-form is to learn facts and design the searching procedure for accessing these facts during the understanding process. For the text-queries $T^G[typ](C)$ or $T^G[part](C)$ learned knowledge is given as a set of categories derived from the category C. The learned categories are stored as the part of learned knowledge. During the understanding process the text-query $T^G[typ](C)$ is transformed into the procedural-form $P([typ](C))$ that contains an appropriate searching procedure $Find\_IDEN\_o$. The procedure $Find\_IDEN\_o$ is designed during the learning stage and the parameter IDEN such as TypeO or PartO is used to search for the required item.

TypeO	**PartO**
O_T1	O_p1
O_T2	O_p2
O_T3	O_p3

The searching-procedural-form specifies the learning procedures that make it possible to store these facts and searching procedure that make it possible to search for these facts during understanding process. The learning procedure and searching procedure are strictly connected and these procedures specify the mechanism of storing and accessing knowledge and employing the different storing and searching methods that make it possible to optimize the understanding process. For example, learning of the searching-procedural-form $P(o\_in\_X)$ is to select an object o_k such as the river and to learn knowledge at each categorical level represented by the container X^i_j. By selecting categorical level (container) X^1_j all objects in this container are learned $L(X^1_j) \triangleright [r^1_1,...,r^1_k]$, where j=1,...,M and M is a number of containers. By selecting next categorical level (container) X^2_j all objects in this container are learned $L(X^2_j) \triangleright [r^2_1,...,r^2_k]$, where j=1,...,H and H is a number of containers.

The main part of the computational-procedural-form consists of the mathematical model (mathematical formula), algorithm or heuristics. The mathematical model can be given in the form of the simple mathematical formula e.g. the linear equation or the differentiation equation. Algorithm can be any numerical algorithm used to compute a complex task.

Learning of the text-task is to learn the query-form, the basic-form and the procedural-form. For example, the text-task "what is difference between 36323 and 35323", has the query-form "CQ_WhatIs CPrALGEBRA Cbetween CN Cand CN", the basic-form " compute difference a1 and a2" and the procedural-form "Data a1, a2 formula F1: r=a2-a1". The similar task "how much greater than 5291 is 5691", has the query-form "CQ_HowMuch CGmPREDICATE Cthan CN CIs CN", the basic-form "how much greater than a1 is a2" and the same procedural-form "Data a1, a2 formula F1: r=a2-a1". In these tasks the computational-procedural-form is given in the form of mathematical formula.

Examples of the text-task for which solution is given in the algorithmic form are following text-task given with its basic-form and the procedural-form. The text-task "which number is between 6857 and 7013 a.6558 b.6842 c.6901 d.7023" is translated into basic-form "Select number that is between a1 and a2" and into the procedural-form: Data: b(1), ... b(n), find b: a1> b(i)<a2. The task "which number has smallest value 0.069 0.2 0.08 0.101" is translated into the basic-form "Find smallest number a(1), ,a(n)" and into the procedural-form: Data [a(1), ,a(n)] find a: for all a(i), a(i)<=a.

The statistical-text-task similarly like the mathematical-text-task can be formulated by applying only statistical categories and solved by applying the computational-procedural-form. For example, the text-task "what is mean of these numbers 10 2 4 5 6 0 9 9 " is translated into the basic-form "Compute mean a(1), ,a(n)" and into the procedural-form "Data [a(1), ,a(n)] compute mean F1: m=(a(1)+...+a(n))/n".

The tasks used in intelligence tests are called the intelligence-text-task. These tasks are usually easy to understand but the solution cannot be found by applying the simple rule or formulae. The intelligence-text-task is solved by applying different heuristics. For example, the intelligence-text-task such as "what is next number in this pattern 29 26 23 20 17", or "which number is missing from this number sequence 20 23 26 ? 32" are the tasks that are solved by applying heuristics. The task "what is next number in this pattern "29 26 23 20 17" is translated into basic-form "Find next number a(1), ,a(n)" and into the procedural-form: Data [a(1), ,a(n)] find a(n+1) by using heuristics H1: r(i)=a(i+1)-a(i), if r(1)= r(2)=...=r(n-1) then a(n+1)=a(n)+r H2: d(i)=a(i+1)/a(i) if d(1)= d(2)=...=d(n-1) then a(n+1)=a(n)+d, ..., Hn. The new heuristics can be added so SUS can solve very different tasks of this type.

During learning stage the query-form for each task is learned along with the basic-form and procedural-form. During understanding process the procedural-form for the computational-text-task describes steps that lead to the solution. These steps are usually given in the following form: the data are extracted, the proper formula is found and the computations are performed. For example, the procedural-form for the computational-text-task is learned following the schema: Data in units, Formula. After computation the result is given by the final form. For the complex text-task the procedural-form describes the detail steps of computation in the form of the concise description or algorithm. For example, for the physical-text-task "auto travels at rate of 25 km/h for 4.0 minutes and finally at 20 km/h for 2.0 minutes find total distance covered in km" the procedural-form is given in the form of the concise description: Data v1 [vU], t1 [tU] v2 [vU], t2 [tU]; Find: s Defining equation: v=s/t, Convert units: v[vU]=v[m/s] t[tU]=t[s] Total distance equals the sum of the distances covered during period t1 and t2, so s=s1+s2=v1*t1+v2*t2.

In this section the procedural-form was described. The procedural-form makes it possible to find a solution to the text-task. However, in order to understand the text-task there is a need to understand meaning of words to be able to understand the text in terms of the real word scene which the text-task usually depicts. These problems will be described in the next section that will present the short introduction to the explanatory script.

7.3.2.9 Understanding and Learning of the Knowledge of the Text-Task – The Explanatory Script

In this section the short introduction to the explanatory script is presented. Problem of the interpretation of the text in terms of the real world situation is connected with problem of understanding of the dictionary-text and it will be presented in the context of the interpretation of the meaning of the dictionary-text. The explanatory script $J(T^T)$ gives the interpretation of the text-task in terms of the real world objects. The explanatory script $J(T^T)$ refers to the meaning of the text-task. The meaning of the text-task is given by both the explanatory script and the basic-form $\beta(T^T) \triangleright (J,B)$. For example, the text-task T0_Buy "Kate bought four apples for \$ 2 each what was total cost" refers to the scenario of "buying" and describes the different aspects of "buying". The interpretation based on the explanatory script is needed to explain, for example, why some information included in the text is redundant and not useful in solving the text-task. In the case of the text-task T0_Buy the information such as name of the subject "Kate" or name of the objects that are bought "apple", is redundant. During learning stage, at the intermediate level used to identify the basic-form, these terms are exchanged by letters "A bought four O for $ 2 each what was total cost". In this type of text-task the information about "who is buying" is not relevant so the text-task is transformed into form "four O for $ 2 each (were bought) what was total cost". This form is transformed into the basic-form "O cost n $ Compute cost N O". The explanatory script identifies the categories that are irrelevant to the process of solving task. The explanatory script of "buying_type_0" is given as "Kate bought 4 apples Each apple cost 2 $ Total cost is C $ Kate need to pay C $ Kate has enough money to pay Kate spend C $ Kate has 4 apples Kate has less money". Based on this script the explanatory form is formulated: Subject: Kate; Object: apple; Action: buy. The explanatory form is used to identify the irrelevant information and to explain why this information is not needed in the process of finding of the solution.

7.3.3 Understanding and Learning of the Knowledge of the Text Object – The Dictionary-Text

In this section the short introduction to the dictionary-text is presented. The dictionary-text T^D is a short text that can be found in the dictionary or encyclopedia. The dictionary text T^D is usually given in the form of a dictionary definition. The encyclopedic or dictionary definitions have the specific form as described in [118]. The short definition is given in the form of one or two sentences. The dictionary-text given in the form of the encyclopedic or dictionary definition will be called the definition-dictionary-text $T^D[Def]$. The definition-dictionary-text $T^D[Def]$ has no more than one sentence. The text "robotics is the branch of technology that deals with the design, construction, operation, and application of robots" is an example of the definition-dictionary-text. The dictionary text that contains biographical information will be called the biographical-dictionary-text $T^D[Big]$. The text "Lippmann Gabriel 1845-1921. French doctor who invented the direct colour process in photography" is an example of the biographical-dictionary-text. The dictionary text that contains knowledge from the domains of the scientific discipline will be named after a particular domain. For example, the dictionary text that contains knowledge from physics will be called the physical-dictionary-text $T^D[Phy]$ e.g. "acceleration is the rate of increase in the velocity of a moving body", "accelerator is apparatus for measuring acceleration". In contrast to short definitions, the long definitions consist of more than two sentences. The dictionary text that contains knowledge from chemistry is called the chemistry-dictionary-text $T^D[Chem]$. For example, the sentences "Acid a substance which in solution in an ionizing solvent (usually water) gives rise to hydrogen ions (H+ or protons). Acids react with alkalis to form salts, and they act as solvent. Strong acid are corrosive; dilute acids have a sour or sharp taste" are good illustration of the chemistry-dictionary-text consisting of more than one sentence.

The meaning of the dictionary texts such as the biographical dictionary texts refer to the facts that can be understood in the context of the interpretational script. These texts do not ask for finding of the unique solution. In contrast, the text-tasks are texts meaning of which refers to the unique solution.

The short texts such as [119] used in helping of the developing skills in reading or testing reading skills that refer to use of commonsense knowledge in description of the real world situation such as "The little pig saw the open gate and ran as fast as its little legs could carry it, away from the farmer, away from its mother and across the paddock. The farmer chased the little pig, across the paddock, through the trees, around the rocks, over the bridge and into a corn field. The little pig stopped to eat some corn and the farmer stopped to take the rest. The little pig stopped eating to take a rest. The farmer scooped up the tired little pig and carried it back to the farm" is called the literature-dictionary-text $T^D[Lit]$. The literature-dictionary-text is much more difficult to understand by SUS than the text that has more formal style such as the biological-dictionary-text $T^D[Bio]$, "The body of the octopus is like a bag with eight long arms. Each arm is lined with suckers. The octopus lives alone in caves or beneath rocks on the ocean floor. It crawls around on the ocean floor looking for food such as shellfish. Its worst enemy is the moray eel. If the octopus is losing of fight with an eel the octopus can release a squirt of black ink to hide behind while it escapes. If it loses an arm in the fight the arm can grow back. "

Understanding of the dictionary-text T^D requires that SUS understands the text content, is able to explain its meaning and is able to answer questions that refer to the meaning of the text. Understanding of the dictionary-text is connected with translation, asking questions, answering questions, giving explanation, learning performing actions, or learning definitions. Understanding of the dictionary text is based on the previously learned explanatory script J. The script includes the descriptions of the stereotypical situations and is given at the different levels of description that reveals different levels of details needed for understanding of the dictionary text. Understanding of the dictionary text at the basic level is to find if the dictionary text interpreted as the logical sentence is true or false. It is similar to answering the educational test where some items have the form "A is Q that has p: true, false" e.g. "Triangle is polygon that has three vertices: true, false". Understanding of the dictionary text at the basic level is to compare the facts of the examined dictionary text with the facts of the learned script. For example, the geometrical-dictionary-text "Triangle is polygon that has three vertices" is transformed into the basic-form "CPolX has 3 CPolPart" and is compared with the fact in the learned interpretational script "triangle 3 vertices". During understanding process the Geometrical-Dictionary Script-Expert invokes the Procedural-Form-Dictionary-Expert that implements the procedural-form of the dictionary text.

The Procedural-Form-Dictionary-Expert is checking if the fact in the basic-form "CPolX has 3 CPolPart", that for the dictionary text "Triangle is polygon that has three vertices" is transformed into the form "triangle 3 vertices" is equal to the fact contained in the learned script "triangle 3 vertices". The Geometrical-Dictionary-Script-Expert and the Procedural-Form-Dictionary-Expert are part of the SUS. SUS is a system consisting of learning, processing, reasoning and decision making experts that communicate between each other during learning and understanding process that is described in more details in Chapter 8.

The mineralogical-dictionary-text $T^D[Min]$ is the short text that could be found in mineralogical books, journals, encyclopedias, dictionaries or magazines. The mineralogical-dictionary-text $T^D[Min]$ has specific form and includes categories of the mineralogical knowledge object. Examples of the mineralogical texts that have different styles are shown below:

TEXT 1 : Cuprite

General Cuprite Information
Chemical Formula: Cu2O
Composition: Molecular Weight = 143.09 gm
Copper 88.82 % Cu 100.00 % Cu2O Oxygen 11.18 % O
100.00 % 100.00 % = TOTAL OXIDE
Empirical Formula: Cu2O
Environment: Oxidized zone of copper deposits.
IMA Status: Valid Species (Pre-IMA) 1845
Locality: Commonly in the copper deposits of SW USA and in Chile. Link to MinDat.org Location Data.
Name Origin: From the Latin, cuprum, meaning copper. Chalcotrichite from the Greek, meaning "hairy copper."
Name Pronunciation: Cuprite
Synonym: Chalcotrichite, ICSD 63281, PDF 5-667, Red oxide of copper
Cuprite Image:
Cuprite Crystallography
Cell Dimensions: a = 4.2696, Z = 2; V = 77.83 Den(Calc)= 6.11
Crystal System: Isometric - HexoctahedralH-M Symbol (4/m 3 2/m) Space Group: P n3m
X Ray Diffraction: By Intensity(I/Io): 2.465(1), 2.135(0.37), 1.51(0.27),
Crystal Structure:
Physical Properties of Cuprite
Cleavage: {111} Imperfect
Color: Brown red, Purple red, Red, Black.
Density: 6.1
Diaphaneity: Transparent to translucent
Fracture: Brittle - Conchoidal - Very brittle fracture producing small, conchoidal fragments.
Habit: Capillary - Very slender and long, like a thread or hair (e.g. millerite).
Habit: Massive - Granular - Common texture observed in granite and other igneous rock.
Hardness: 3.5-4 - Copper Penny-Fluorite
Luminescence: Non-fluorescent.
Luster: Adamantine
Magnetism: Nonmagnetic
Streak: brownish red
Optical Properties of Cuprite
Gladstone-Dale:

CI meas= -0.296 (Poor) - where the CI = (1-KPDmeas/KC)
CI calc= -0.294 (Poor) - where the CI = (1-KPDcalc/KC)
KPDcalc=0.3028,KPDmeas=0.3033,KC=0.234 Ncalc = 2.43
Optical Data: Isotropic, n=2.85.
RL Color: Gray blue with internal red reflections.
RL Pleochroism: Anomalous.
Reflectivity
Standardized Intensity (100%) Reflection Spectra of Cuprite in Air
Calculated Properties of Cuprite
Electron Density: Bulk Density (Electron Density)=5.64 gm/cc note: Specific Gravity of Cuprite =6.11 gm/cc.
Fermion Index:
Fermion Index = 0.00005367 Boson Index = 0.99994633
Photoelectric: PECuprite = 40.46 barns/electron
U=PECuprite x Electron Density= 228.05 barns/cc.
Radioactivity: GRapi = 0 (Gamma Ray American Petroleum Institute Units) Cuprite is Not Radioactive
Cuprite Classification
Dana Class:
04.01.01.01 (04)Simple Oxides
(04.01)with a cation charge of 1+ (A+2 O)
(04.01.01)Dana Group
04.01.01.01 Cuprite Cu2O P n3m 4/m 3 2/m
Strunz Class:
04.AA.10 04 - OXIDES (Hydroxides, V[5,6] vanadates, arsenites, antimonites, bismuthites, sulfites, selenites, tellurites, iodate
04.A - Metal:Oxygen = 2.1 and 1:1
04.AA_-Cation:Anion (M:O) = 2:1 (and 1.8:1)
04.AA.10 Cuprite Cu2O P n3m 4/m 3 2/m

TEXT 2 [120]:
Crystals are octahedral, cubic, and dodecahedral; twining is uncommon. Cuprite also occurs in massive, compact, and granular habits. The colour ised, and the streak aownish red. Cuprite is a translucent to transparent mineral. When exposed to the air, it tarnishes to semi-opaque. It has an adamantite, submetalic or earthy lustre. Formation This widespread mineral forms in the oxidized parts of copper deposits, where it is associated with native copper, malachite, azurite, chalcocine, and oxides of iron. Tests It is soluble in nitric and other acids. It fuses, turning the flame green

TEXT 3 [121]:
Cuprite Oxides and hydroxides Cu2O (coper oxide) System Isometric. Appearance Octahedral, cubic or dodecahedral crystals or a combination of the above forms. Dark ruby-red color, sometimes altered on the surphace to green malachite. Agregates of fine, acicular, hairlike crystals (chalcotrichite variety) occur. Physical properties Semi-hard (3.5-4), very heavy, fragile, poor cleavage. Translucent with adamantine luster. Altering to semi-opaque if exposed to air. Bright red streak. Soluble in concentrated acids. Fuses readily, turning the flame green (copper). Environment found in the osidation zone of copper deposits associated withazurite, tenorite, malachite and native copper. Occurrence Fine octahedral crystals at Chessy (France), Redruth and Liskeard (Cornwell, England), Bisbee and Morenzi, Arizona (USA) Corocoro (Bolivia), Chuquicamata (Chile), Onganja (Namibia); deposits in the Urals and Altai (USSR) and on Sardinia. Halcotrichite is found at Morenzi, Arizona (USA), Monsol (France), in Chile and at Libiola (Liguria, Italy). Uses Important ore of coper. Some flawless, transparent crystals have been cut as gems.

TEXT 4 [122]:
Cuprite A major ore of coppe, cuprite is named from the Latin cuprum meaning "coper". It can turn superficially dark grey on exposure to light. Cuprite typically has cubic crystals. In the

variety called chalcotrichite or plush copper ore, the crystals are fibrous and found in loosely matted aggregates. Cuprite is a secondary mineral, formed by the oxidation of copper sulphide veins. Fine specimen come from Namibia, Australia, Russia, France and the USA.

TEXT 5 [123]:

Anhydrite

General Category Mineral

Chemical formula anhydrous calcium sulfate:CaSO4

Identification Color [Colorless, White, Bluish white, Violet, Dark gray]

Crystal habit [very rare tabular and prismatic crystals. Usually occurs as fibrous, parallel veins that break off into cleavage fragments. Also occurs as grainy, massive, or nodular masses]

Crystal system Orthorhombic; 2/m 2/m 2/m

Cleavage [010] Perfect, [100] Perfect, [001] Good

Fracture conchoidal

Mohs Scale hardness 3.5

Luster Vitreous - Pearly

Refractive index n?=1.569 - 1.573 n?=1.574 - 1.579 n?=1.609 - 1.618

Pleochroism Biaxial (+)

Streak white

Specific gravity 2.97

Fusibility 2

Other Characteristics

Some specimens fluoresce; many more fluoresce after heating

Anhydrite is a mineral - anhydrous calcium sulfate, CaSO4. It is in the orthorhombic crystal system, with three directions of perfect cleavage parallel to the three planes of symmetry. It is not isomorphous with the orthorhombic barium (barite) and strontium (celestine) sulfates, as might be expected from the chemical formulas. Distinctly developed crystals are somewhat rare, the mineral usually presenting the form of cleavage masses. The hardness is 3.5 and the specific gravity 2.9. The colour is white, sometimes greyish, bluish or purple. On the best developed of the three cleavages the lustre is pearly, on other surfaces it is vitreous.

When exposed to water, anhydrite readily transforms to the more commonly occurring gypsum, (CaSO4•2H2O) by the absorption of water.

Anhydrite is commonly associated with calcite, halite, and sulfides such as galena, chalcopyrite, molybdenite, and pyrite in vein deposits.

Anhydrite is most frequently found in salt deposits with gypsum; it was, for instance, first discovered, in 1794, in a salt mine near Hall in Tirol. In this occurrence depth is critical since nearer the surface anhydrite has been altered to gypsum by absorption of circulating ground water.

From an aqueous solution calcium sulfate is deposited as crystals of gypsum, but when the solution contains an excess of sodium or potassium chloride anhydrite is deposited if temperature is above 40°C. This is one of the several methods by which the mineral has been prepared artificially, and is identical with its mode of origin in nature, the mineral is common in salt basins.

The name anhydrite was given by A. G. Werner in 1804, because of the absence of water of crystallization, as contrasted with the presence of water in gypsum. Some obsolete names for the species are muriacite and karstenite; the former, an earlier name, being given under the impression that the substance was a chloride (muriate). A peculiar variety occurring as contorted concretionary masses is known as tripe-stone, and a scaly granular variety, from Vulpino, near Bergamo, in Lombardy, as vulpinite; the latter is cut and polished for ornamental purposes.

Chicken-wire anhydrite

Chicken-wire anhydrite is a type of anhydrite deposit where seeping ground water gradually deposited large amounts of anhydrite in the sediment, replacing most of it, so that a cross-section of it looks somewhat like the coarse wire netting often used to confine poultry.

Text 1 is has a most formal style and can be easily understood. Text 2, Text 3 and Text 4 gives the description of minerals in the form of the mineralogical categories that refers to the interpretational script. Text 5 has both formal part and description in paraformal style.

Understanding of the mineralogical-dictionary-text $T^D[Min]$ requires transforming it into the query-form and into the interpretational script J . The mineralogical-dictionary-text $T^D[Min]$ is presented in the context of the system of the minerals understanding "Domeyko" described in Section 7.2. Understanding of the mineralogical-dictionary-text is based on the model of the mineralogical objects. The model of the mineralogical objects represents knowledge of the minerals category v_M^i and is used to interpret the mineralogical-dictionary-text $T^D[Min]$ as the part of the interpretational script and to supply knowledge during understanding of the objects from the mineralogical category. The model of the mineralogical objects that represents knowledge of the mineral category v_M^i is part of the mineralogical knowledge (knowledge of minerals categories) $\kappa_{KOb}[Min]$. Knowledge of the minerals categories $\kappa_{KOb}[Min]$ is gathered in the process of scientific mineralogical research. For each mineral category v_M^i knowledge in the form of the mineralogical description of the selected mineral category is gathered based on the complex process of scientific investigation.

Both, the empirical finding and theoretical description in the form of definitions are used to build the model of the mineral object. Each mineral category v_M^i is described by a set of attributes a_j^i with value of attribute denoted as $val(a_j^i)$. The physical attributes that are characteristic for all real world objects are denoted by a_j^{Ph}. The category of physical attributes include attributes such as the size $a(L)_j^{Ph}$, the specific gravity $a(G)_j^{Ph}$, the hardness $a(H)_j^{Ph}$, the color $a(C)_j^{Ph}$, the streak $a(S)_j^{Ph}$, the cleavage $a(K)_j^{Ph}$, the diaphaneity $a(D)_j^{Ph}$, the luster $a(L)_j^{Ph}$, the habit $a(H)_j^{Ph}$, the tenacity $a(T)_j^{Ph}$, the fluorescence $a(F)_j^{Ph}$, the magnetism $a(M)_j^{Ph}$, the radioactivity $a(R)_j^{Ph}$, the piezoelectricity $a(P)_j^{Ph}$. These physical attributes have usually the specific meaning in area of minerals' recognition. Each qualitative physical attribute has its value given as the attributes category. The value of the i-th attribute for the j-th mineral is denoted by $val(a_j^i)=\{c_1^i,...,c_K^i\}$, where K is a number of categorical data for j-th mineral, whereas the values of the i-th attribute for all minerals give the domain of the i-th attribute $val(a^i)=\langle c_1^i,...,c_K^i\rangle$, where M is a number of the attribute categories. The quantitative physical attribute has its value given as the numerical value, discrete or continuous. A discrete attribute is a quantitative attribute whose value is countable. The hardness $a(H)_j^{Ph}$ is a discrete attribute that has value $val(a(H)_j^{Ph})=\{1,...,10\}$. The magnetism $a(M)_j^{Ph}$ is a discrete attribute that has value $val(a(M)_j^{Ph})=\{0,1\}$. The specific gravity $a(G)_j^{Ph}$ is a continuous attribute that has value in the interval [1-20]. The color $a(C)_j^{Ph}$, the streak $a(S)_j^{Ph}$, the cleavage $a(K)_j^{Ph}$, the diaphaneity $a(D)_j^{Ph}$, the luster $a(L)_j^{Ph}$, the habit $a(H)_j^{Ph}$ are categorical attributes whose values are attribute categories c_k^j. Color $a(C)_j^{Ph}$ has its values as a set of color categories $val\left(a(C)_j^{Ph}\right) \equiv \{c_1^C,...,c_N^C\}$. Because the color categories $\{c_i^C\}$ are established based on the subjective criterion they can be difficult to match during the process of recognition and understanding. In order to avoid an error during matching process these categories are transformed into the basic color categories $\{c_i^{CB}\}$. The habit attribute $a(H)_j^{Ph}$ is divided into the crystal face attribute $a(Hf)_j^{Ph}$, the crystal form attribute $a(Hc)_j^{Ph}$ and the crystal aggregate attribute $a(Ha)_j^{Ph}$. Similarly like the color, the habit attribute $a(H)_j^{Ph}$ has its values as a set of habit categories $val\left(a(H)_j^{Ph}\right) \equiv \{c_1^H,...,c_N^H\}$. For example, the crystal face attribute $a(Hf)_j^{Ph}$ has its values $val\left(a(Hf)_j^{Ph}\right) \equiv \{c_1^{Hf},...,c_N^{Hf}\}$.

The chemical attributes a_j^{Ch} are attributes connected with chemical properties of minerals. The chemical attribute a_j^{Ch} includes: the reactivity to dilute acids $a\left(D^A\right)_j^{Ch}$, the reactivity to dilute water $a\left(D^W\right)_j^{Ch}$, the flame test attribute $a(F\ )_j^{Ch}$, the instrumental method spectroscope attribute $a\left(I^S\right)_j^{Ch}$, the instrumental method infrared spectrograph attribute $a\left(I^{IR}\right)_j^{Ch}$, the instrumental method of the differential thermal analysis attribute $a\left(I^{DITA}\right)_j^{Ch}$.

The optical attributes a_i^P are given in the form of the characteristic features such as type $a(T)_i^P$, RI values $a(R)_i^P$, surface relief $a(S)_i^P$, dispersion $a(D)_i^P$, Gladstone-Dale index $a(G)_i^P$ or pleochroism $a(P)_i^P$. The X Ray Diffraction analysis attributes a_i^X, are attributes connected with X Ray Diffraction analysis and include attributes such as X-Ray Powder Diffraction attribute $a(P)_i^X$, value $a(P)_i^X$ is given as a set of numbers [Radiation, d-spacing, intensity].

All minerals have distinct crystal structures. Atoms which make up mineral are arranged in regular and repeating three dimensional network. The crystal structure s_i is represented by the crystal system and 230 space groups. The crystal structure is given by two characteristic attributes Z, W $s_i\left[s_i^Z, s_i^W\right]$. The symmetry attribute s_i^W is expressed by the symbol for a space group that consists of the centering followed by an abbreviated Hermann-Mauguin symbol. For example, symbol $P2_12_12_1$ is a primitive unit cell in the orthorhombic system in which the twofold axes are twofold screw axes and the symbol $F4/m\bar{3}2/m$ is a face centered unit cell in the cubic system. The cell attribute parameters s_i^Z is represented by the axial ratios $s(a)_i^Z$, the cell dimensions $s(d)_i^Z$, the unit cell $s(u)_i^Z$ and the volume $s(v)_i^Z$. Knowledge of crystal structure is expressed as $s_i\left[s_i^Z\left[s(a)_i^Z, s(d)_i^Z, s(u)_i^Z, s(v)_i^Z\right], s_i^W\right]$ and knowledge of the mineral is expresed as $v_M^i \equiv \varphi_i\left[\varphi_i^N, \varphi_i^F\right]s_i\left[s_i^Z, s_i^W\right]a_i\left[a_j^{Ph}, a_j^{Ch}, a_j^P, a_j^X\right]$.

Mineral has distinctive chemical composition φ_i which may vary within certain limit but is always clearly defined. Chemical composition is given in the form of the name $\varphi(N)_i^{ChC}$ and the symbolic formula $\varphi(F)_i^{ChC}$ e.g [copper carbonate hydroxide, CuSO4], essential elements φ^E and trace elements φ^T: $\varphi_i\left[\varphi_i^{ChC}\left[\varphi(N)_i^{ChC}, \varphi(F)_i^{ChC}\right], \varphi^E, \varphi^T\right]$. Formula can be given in the form of patterns A, A_mB_n, $A_mB_nC_w$, $A_mB_nC_wD_v$ where A, B, C, D are chemical elements. Formula can be also given in the form of the specific anionic groups A_mB_n, $A_mB_nX_w$. The chemical formula can be given as formula of the solid solution $(A_m, B_n)X_w$. Based on the type of anionic group the different categories of the mineral groups are derived from the mineral category. The category of groups of minerals is divided into the native elements category, the sulfides category, the sulfosalts category, the oxides category, the hydroxides category, the halides category, the carbonates category, the nitrates category, the borates category, the sulfates category, the phosphates category, and the silicates category and is represented by the categorical chain as follows: $\triangleright\left\langle v_{Min}\right\rangle \triangleright\left\langle v_{Nel}, v_{Sul}, v_{Sfo}, v_{Oxi}, v_{Hyd}, v_{Hal}, v_{Sfo}, v_{Car},\right\rangle$. For example, the category of oxides group that is derived from the mineral category is given by specifying the chemical composition φ_i^{ChC} of an anion group A $\varphi\{A\}_i^{ChC}$ as $\varphi\{A = O\}_i^{ChC}$, where O means oxygen. The knowledge of the mineralogical category includes only limited knowledge of chemical facts. The mineralogical interpretational script stores chemical knowledge for anion group $\varphi\{A = O\}_i^{ChC}$, in which the chemical composition φ_i^{ChC} is given by the chemical formula $\varphi_i^{ChC}\left[\varphi(N)_i^{ChC}\{a = O\}, \varphi(F)_i^{ChC}\{a = O\}\right]$ and by two components: the name N and symbolic formula F. For example, mineral sylvite is given as "sylvite is potassium chloride KCl". In order to understand the meaning of the terms "potassium chloride KCl" the script from the chemical knowledge category $\kappa\,[Che]$ is needed.

Each mineral is given by its name χ_i^N and a special index ∂_i. The name χ_i^N is expressed in one of existing languages L_j such as English $\chi_i^N(E)$, or Polish $\chi_i^N(P)$. The name is also connected with the synonyms that are given in one of the known languages such as English $\chi_i^N(E)$. The special index ∂_i refers to one of the index systems such as the system of Nickel-Strunz symbols ∂_i^{N-S}, the Hey's

Index ∂_i^H or the Dana index ∂_i^D and can be written as $\partial_i\left[\partial_i^{N-S},\partial_i^H,\partial_i^D\right]$. A mineral is often represented by the index, name, chemical formula and symmetry attribute as shown in the tabele:

01.AA.05	Aluminum	Al	F m3m, P m3m	4/m 3 2/m
01.AA.10a	Auricupride	Cu3Au	P m3m	4/m 3 2/m

One important part of mineralogical knowledge is IMA status $\hbar_i$ and name of origin Λ_i. The IMA status $\hbar_i$ is given by the year of approval $\hbar_i^Y$. Knowledge of the locality λ_i where the mineral occurs is very useful information in the process of mineral recognition. The locality is referring to geographical knowledge and usually it includes a country λ_i^C, a province λ_i^P, a place λ_i^L, a mine λ_i^M and is expressed as. $\lambda_i\left[\lambda_i^C,\lambda_i^P,\lambda_i^L,\lambda_i^M\right]$ The example of the description of the locality is as follows: White Queen Mine, Hiriart Mountain, Pala District, San Diego Co., California, USA. The mineral category is part of the rock category $..\triangleright\left\langle v_{Rock}\right\rangle\succ[v_{Min}]$. The part of mineralogical knowledge is concerned with the geological setting γ_i and associated minerals θ_i. The name of the origin Λ_i is given by the year of naming Λ_i^Y, the name of the discoverer Λ_i^N the name of the source Λ_i^S and can be written as $\Lambda_i[\Lambda_i^Y,\Lambda_i^N,\Lambda_i^S]$. A mineral can be named Γ_i for the name of professional Γ_i^N, appearance Γ_i^A and can be written as $\Gamma_i[\Gamma_i^N,\Gamma_i^A]$.

The model of the mineral object that is part of the knowledge of the mineral category $\kappa\left[v_M^i\right]$ and that incorporates basic mineral data can be written as:

$$v_M^i\equiv\varphi_i[\varphi_i^{hC},\varphi_i^F,\varphi_i^T]s_i[s_i^Z,s_i^W]a_i[a_j^{Ph},a_j^{Ch},a_j^P,a_j^X]\chi_i[\chi_i^N(L),\chi_i^S(L)]\partial_i[\partial_i^{N-S},\partial_i^P,\partial_i^H]\hbar_i[\hbar_i^Y]\lambda_i[\lambda_i^C,\lambda_i^P,\lambda_i^L,\lambda_i^M]\Lambda_i[\Lambda_i^Y,\Lambda_i^N,\Lambda_i^S]\Gamma_i[\Gamma_i^N,\Gamma_i^A]\gamma_i\theta_i$$

The model of the mineral object is used as a part of the interpretational script implemented as the Procedural-Form-Dictionary-Expert. During understanding the Mineral-Dictionary-Script-Expert invokes the Procedural-Form-Dictionary Expert to interpret the meaning of the text.

The simple mineralogical dictionary text such as "malachite has green color" is translated by the Mineral-Dictionary-Script-Expert into the basic-form by invoking the Procedural-Form-Dictionary-Expert that implements the interpretational script "[CatMineral is CatRealWOobject] [CatMineral has CatMinAttribute] [CatMineral has CatMinAtrColor CatValMinAtrColor] [CatMineral is CatValMinAtrColor=CatMineral has CatMinAtrColor CatValMinAtrColor]". Based on this script the text is translated into basic-form "Mineral (malachite) Color (green)". The text "malachite has green color" is easy to translate because all categories refer to the categories of script. The text "malachite is green" is translated based on the script [CatMineral is CatValMinAtrColor=CatMineral has CatMinAtrColor CatValMinAtrColor] into the basic-form "Mineral (malachite) Color (green)". This script supply additional information that word "green" is member of the CatValMinColor category and this category refers to attribute category CatMinColor.

Both of these texts have the same meaning. The Mineral-Dictionary-Script-Expert invokes the Procedural-Form-Dictionary-Expert that compares the basic-form "Mineral (malachite) Color (green)" with the learned basic-form "Mineral (malachite) Color [green]". It makes it possible to find if the text "malachite has green color" give true information by comparison the text translated into basic-form "Mineral (malachite) Color (green)" with learned procedural-form "Mineral (malachite) Color [green]". If the result of comparison by the Procedural-Form-Dictionary-Expert is "true" it means that the mineral named malachite, described in the text, has the value of colour attribute "green", that is in agreement with knowledge that was previously learned. The learned basic-form is expressed in the more formal way as $val(a(C)_j^{Ph}) \equiv \{c_1^C,...,c_N^C\}(v)$, that means mineral "j" has N categorical values $\{c_1^C,...,c_N^C\}$ of the attribute $val(a(C)_j^{Ph})$. The basic-form obtained during translation is given as $val(a(C)_j^{Ph}) \equiv \{x_1^C,...,x_N^C\}(v)$. Basic understanding of the dictionary text is to find the result of comparison $\{c_1^C,...,c_N^C\}$ and $\{x_1^C,...,x_N^C\}$. If $\{c_1^C,...,c_N^C\} \equiv \{x_1^C,...,x_N^C\}$ then system understands text. This can be also interpreted as answering question "malachite has green color: true or false?". The simple understanding means to understand the text at the basic level. Complex understanding, understanding at the higher level is to generate, as the result of understanding, the complex conformation of the text. For example, "Malachite is green but its streak is pale green", "malachite is green but it is not so hard as turquoise", "malachite copper carbonate hydroxide CuSO4 is green because most Cu composite are green". The text "Malachite is green but its streak is pale green" is generated by referring to the basic-form $val(a(C)_j^{Ph}) \equiv \{c_1^C,...,c_N^C\}(v_1)val(a(S)_j^{Ph}) \equiv \{c_1^S,...,c_N^S\}(v_2)$. Similarly the text "Malachite is green and its streak is pale green" is understood by referring to basic-form $val(a(C)_j^{Ph}) \equiv \{c_1^C,...,c_N^C\}(v_1)val(a(S)_j^{Ph}) \equiv \{c_1^S,...,c_N^S\}(v_2)$. The basic-form that is part of the script can be used to understood the text that consist of values of more than two attributes such as "Malachite is green its streak is pale green and hardness is 4" that is given as $val(a(C)_j^{Ph}) \equiv \{c_1^C,...,c_N^C\}(v_1)val(a(S)_j^{Ph}) \equiv \{c_1^S,...,c_N^S\}(v_2)val(a(H)_j^{Ph}) \equiv \{y\}(v_3)$ or in general form $\{c_1^1,...,c_N^1\}... \wedge \{c_1^K,...,c_N^K\}$, where K is a number of predicates.

The script utilize the knowledge of the categorical chain. Based on the knowledge of the categorical chain the information that the mineral category is derived from the category of non-living object $\langle v_{Lliv} \rangle$ and non-man made object $\langle v_{Min} \rangle$ can be incorporated into the mineral model. Also the right part of the categorical chain $\triangleright \langle v_{Min} \rangle \triangleright \langle v_{Nel}, v_{Sul}, v_{Sfo}, v_{Oxi}, v_{Hyd}, v_{Hal}, v_{Sfo}, v_{Car}, \rangle$ can be used to build the mineral model that has the chemical composition and belongs to one of the mineral groups.

For example, "franklinite is a zinc manganese iron oxide, and a member of the spinel group" is understood based on the knowledge of the categorical chain.

Based on the categorical chain $\rhd\langle v_{Min}\rangle\rhd\langle v_{Oxi}\rangle\rhd\langle v_{Spin}\rangle\rhd\langle v_{Fran}\rangle$ it is easy to find that franklinite is a member of the spinel group and the spinel group is derived from the oxides group. The spinel group shares some properties that are common for all minerals that are members of this group. The text "franklinite is a zinc manganese iron oxide, and a member of the spinel group" is translated into script "mineral is member of group $M \in \{G_1 \to G_2\}$ has chemical formula $M(F)$ whereas the text "magnetite is iron oxide in spinel group of minerals and has colour black" is translated into script "mineral is member of group $M \in \{G_1 \to G_2\}$ has chemical formula $M(F)$ and minerals predicate $M(P_C)$ has value $M(F) \in M \in \{G_1 \to G_2\} \wedge M(P_c) = [c_1^1,...,c_N^1]$. This script can be given in a more general form: $M(F) \in M \in \{G_1 \to G_2\} \wedge \{M(P_c) = [c_1^1,...,c_N^1]... \wedge M(P_K) = [c_1^K,...,c_N^K]\}$.

Understanding is to find access to learned knowledge in the form of definition. The definition is part of the knowledge of the knowledge schema. The mineral definition is given by the script $\Im^H_{CH,S} \equiv \{M\} \rhd [O](haz)\{(Ch)(S)\}$ that means (mineral M is non man-made object O that has chemical composition $\{Ch\}$ and crystal structure (S). The category of oxides group that is derived from the mineral category is given by specifying of the chemical composition $\{Ch\}$ as the anion group A $\{Ch(A)\}$. The category of the chemical composition $\{Ch\}$ refers to the category of the chemical knowledge $\langle \kappa_{Ch}\rangle$. The category of the chemical knowledge $\langle \kappa_{Ch}\rangle$ stores chemical knowledge such as knowledge of elements, chemical compositions, or chemical reactions. The category of the mineralogical knowledge has only limited knowledge of chemical facts.

The interpretational script stores chemical knowledge $M\{Ch[N,F]\}$, in which the chemical formula $\{Ch[N,F]\}$ is given by the two components: the name N and the symbolic formula F. For example, mineral sylvite is given as "sylvite is potassium chloride KCl". In order to understand the meaning of the terms "potassium chloride KCl" the script from chemical knowledge category $\langle \kappa_{Ch}\rangle$ is needed. The text "malachite copper carbonate hydroxide CuSO4 is green" is translated into script $M\{Ch[N,F]\} \wedge M(P_C) = [c_1,...,c_N]$, where chemical formula $\{Ch[N,F]\}$ is given in the form of name N and symbolic formula F, and minerals predicate $M(P_C)$ has value $[c_1,...,c_N]$. Based on this script the text "malachite is green" will be interpreted as "malachite is green because it is copper carbonate hydroxide" or "malachite is green because it is copper composite". In this Chapter the interpretation of the mineralogical-dictionary-text based on the model of the mineralogical objects was presented. The more detail description of learning and understanding of the dictionary-texts will be given in Chapter 8.

8 Knowledge Implementation

In Chapter 7 the theoretical framework for the knowledge implementation was presented. The knowledge implementation approach described in chapter 7 is used for learning of the knowledge and skills of the different categories of objects. In this Chapter the knowledge implementation as learning of the knowledge and skills is presented. The knowledge implementation is part of the shape understanding method that is aimed at building the understanding/thinking machine, the shape understanding system (SUS). The short description of the shape understanding system is presented in Section 8.1.1, however in this book problems connected with the implementation of the shape understanding system (SUS) will not be presented. The reason is that the theoretical issues, connected with learning and understanding are very complex, and the attempt to describe the implementation problems could, instead of clarifying things, make them less understandable in the context of the material presented in this book.

In Section 8.1, learning of the new skills will be presented, whereas in Section 8.2, learning of the new knowledge will be presented. Learning of the new skills and new knowledge are processes that are strictly connected. Learning of the different skills to deal with processing of the different types of information is approached by designing and implementing of the different processing algorithms. The implemented algorithms are usually used to solve the very specific processing problem. In Section 8.1 learning of the knowledge of the visual objects and sensory object is presented. Learning of the knowledge of the visual objects is connected with application of the complex image processing methods to extract the perceptual data. This can be seen as learning of the specific perceptual skills to acquire data from an image. In Section 8.1.2 learning of the skills of the visual object is presented. Learning of the knowledge of the visual object can be seen as learning of the specific perceptual skills to acquire data from an image. In this book, instead of, "learning of the new skills to deal with processing of the different data of the category of the visual object" we will write "learning of the knowledge and skills of the visual object". In Section 8.1.2.1 learning of the knowledge and skills of the specific categories derived from the sign category is presented. Knowledge of the specific categories derived from the sign category is learned at the prototype level. In Section 8.1.2.1 learning of the knowledge and skills of the letters category is presented. Learning of the knowledge of the letters category is to learn the visual knowledge of letters from all existing writing systems and the non-visual knowledge connected with using a letter during reading and understanding of a text. In Section 8.1.2.4 learning of the knowledge and skills of the arrow category is presented. In Section 8.1.2.4 learning of the knowledge and skills of the category of road signs is presented and in the section 8.1.2.5 learning

Z. Les & M. Les: Shape Understanding System - Knowledge Implementation and Learning, SCI 425, pp. 123–219.
springerlink.com

of the knowledge and skills of the knife category, as an example of the real world objects category is presented. Learning of the knowledge and skills of the category of the sensory object is presented in Section 8.1.3. Learning of the procedural-form that will be presented in chapter 8.1.4 is connected with learning of the problem solving skills that are used to solve educational tasks. Learning of new problem solving skills is concerned with learning of the new methods of solving of the problem and implementing of new methods of processing and storing of acquired knowledge. Learning of the problem solving skills involves learning of the knowledge and mechanisms how to apply acquired knowledge to obtain the solution of a problem (text-task). Because there are numerous methods to solve problems and the solution methods are usually related to each other so SUS need to learn to solve many different problems for each specific topic.

In Section 8.2, learning of the new knowledge is presented. As it was described, learning of the new skills and new knowledge are processes that are strictly connected. Learning of the knowledge and skills connected with understanding of the text will be called the learning of the new knowledge, whereas learning of the new knowledge and skills connected with understanding of the visual or sensory object will be called the learning of the new skills. Learning of the knowledge of the visual object can be seen as learning of the specific perceptual skills to acquire data from an image and because of this learning of the knowledge of the visual object is presented in the section 8.1 as learning of the new skills. Learning of the new knowledge that is used during understanding of the text requires transferring of the knowledge from the different available sources and means, such as dictionaries or handbooks into SUS. In this book, learning and understanding of the text that belongs to one of the text categories such as the category of text-queries, the category of text-task, the category of dictionary-texts, is presented. Knowledge associated with the text category includes the categorical chain, the knowledge network of the categorical chains, the coding categories, the query-form, the basic-form, the procedural-form, the explanatory script, and the interpretational script. The categories that are basic components of acquired knowledge by SUS will be presented in Section 8.2.1. Learning of the categories of the visual objects represented by the categorical chain will be described in Section 8.2.1.1. Learning of the knowledge categories will be described in Section 8.2.1.2. Learning of the coding categories will be described in Section 8.2.1.3. Learning of the knowledge schema will be described in Section 8.2.2. Learning of the knowledge of the text-query will be described in Section 8.2.3. Learning of the knowledge of the text-task will be described in Section 8.2.4. Learning of the knowledge of the query-form will be described in Section 8.2.4.1. The query-form is the generalization of the text-task expressed in the form of the clauses consisting of the coding categories. In Section 8.2.4.2 learning of the basic-form that refers to the basic meaning of the text-task will be presented. In Section 8.2.4.3 learning of the explanatory script is presented. The mathematical-text-task, the physical-text-task, the geographical-text-task are the different categories of the educational tasks. In Section 8.2.4.4 learning of the mathematical-text-task is presented. In Section 8.2.4.5 learning of the physical-text-task is described. In Section 8.2.4.6 learning of the geographical-text-task is described. In Section 8.2.4.7 learning of the tools-text-task is described. In Section 8.2.5 learning of the dictionary-text is described.

8.1 Knowledge Implementation – Implementation of the New Skills

As it was described, learning of the new skills and new knowledge are processes that are strictly connected. Learning of the new knowledge and skills that is connected with understanding of the visual or sensory object can be seen as learning of the specific perceptual skills to acquire sensory data and will be called learning of the new skills. Learning of the specific perceptual skills do not require to learn complex knowledge in contrast to learning of the interpretational skills needed in understanding of the meaning of the text or solving the text-task problem. The specific perceptual skills are utilized in understanding of the visual or sensory object that is connected with naming of the object. When an object from the category of text objects is understood by finding meaning of the text, an object from the category of visual objects or sensory objects is understood by naming of an examined object. The visual object after naming is interpeted based on knowledge of the ontological visual categories and knowledge of the knowledge scheme. Learning of the different skills to deal with processing of the different types of information is approached by designing and implementation of the different algorithms. The implemented algorithms are usually used to solve the very specific processing problem. The complex problem such as a visual reasoning is solved by implementing many different specific algorithms that are governed by the general algorithm of the visual reasoning. Learning new skills by SUS is to acquire the diverse problem solving skills and different processing and reasoning ability. Learning of the problem solving skills involves learning of the knowledge and mechanisms how to apply acquired knowledge to obtain the solution of a problem (text-task). The important problem solving skills are skills used to solve educational tasks. Learning of the problem solving skills starts with the more basic problems and progresses to those that are most difficult. Learning of the problem solving skills that are used to solve educational tasks is connected with learning of the procedural-form that is described in chapter 8.1.4. The short description of the shape understanding system will be presented in the next Section. The shape understanding system (SUS) is designed to learn new skills and knowledge to be able to understand and to think.

8.1.1 Implementation of the New Skills – Shape Understanding System

In this Section the short description of the shape understanding system is presented, however in this book problems connected with the implementation of the specific parts of SUS will not be discussed. The reason is that the theoretical issues, connected with learning and understanding are very complex, and the attempt to describe the implementation problems could, instead of clarifying things, make them less understandable in the context of the material presented in this book. As it was described, Shape Understanding System (SUS) is a system consisting of learning, processing, reasoning and decision making experts that communicate between each other during learning and understanding process. In this book, the issue connected with implementation of experts is presented only in the context of the general description of the system. The names of the experts applied throughout the book are given to show that all processing methods or different forms of the knowledge representation implemented in SUS are the part of an expert.

The system may cooperate with distributed experts by utilizing DCOM (Distributed Component Object Model) technology [124] to have access to expertise from an expert that can perform a very specific task such as text understanding and that is part of another system. SUS is built to support the communication in the form understandable for the human user (natural language sentences, icons, signs, images). Experts are abstractions that encapsulate the properties and behaviour of the entities within a system. The expert is epistemologically oriented, that means it is part of the understanding system, it has intelligent behaviour, it performs part of the specific task in the context of the final goal. The system is being developed and new experts are being implemented and added to the system. Experts used for learning are described in the next chapters.

The central module that consists of the Master-Expert E^M, the Reasoning-Expert E^R, the Manager-Expert E^Q and the Processing-Eexpert E^P operates based on knowledge of image processing, decision making and the search strategies as well as knowledge of shape description and representation distributed among the specialized experts. In this book, experts are described in the context of performing the specific processing or reasoning tasks. In order to avoid confusion caused by presentation of the specific implementation issues, the implementation details are not presented in this book.

Analyzing and designing the system involves combining several object-oriented techniques. The class diagram and sequence diagram, as a part of the UML (Unified Modelling Language), were applied to illustrate the designing stages. To illustrate how experts interact with each other sequence diagrams are used (see Fig.8.1). They focus on message sequence, that is, how messages are sent and received between a numbers of experts. On the horizontal axis are the experts involved in a sequence of actions. Communication between experts is represented as horizontal message lines between the object's lifelines. An example of the sequence diagram shown in Fig. 8.1 illustrates a simple scenario of the reasoning process. The names in the boxes denote the experts of the system (MASTER_E denotes the Master-Expert, REASONING_E denotes the Reasoning-Expert, DM1, DM2 denotes the Decision-Making-Experts, MENAGER_E denotes the Manager_Expert (the Data-Acquisition-Expert), ProcE1 and ProcE2 denote The Processing-Expert and ENDANALYSIS_E denotes the End-Analysis-Expert). Description of the selected aspects of reasoning process is given in the following Section. Examples of an implementation that shows selected parts of the reasoning process (C++ style) are also presented in this chapter.

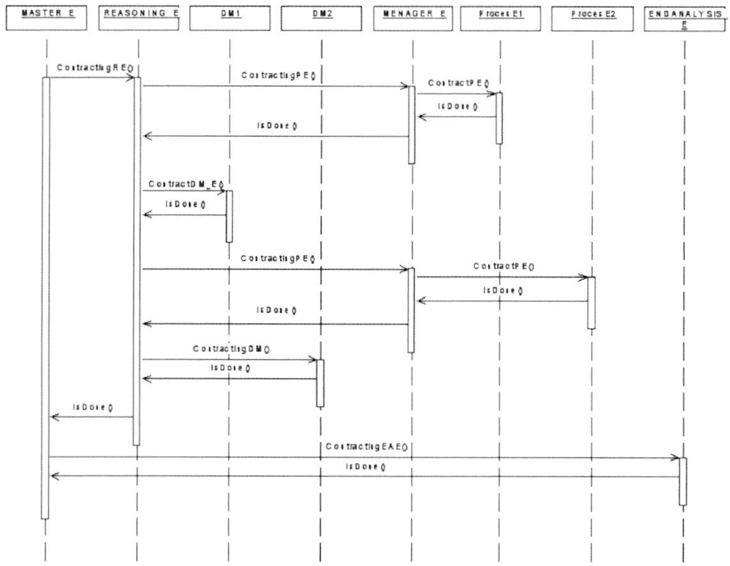

Fig. 8.1. A sequence diagram with one scenario of understanding process

Reasoning-Expert

The Reasoning-Expert E^R is an expert that manages the process of reasoning and is invoked by the Master-Expert E^M. The Reasoning-Expert E^R makes a decision based on the expertise α of the Decision-Making-Expert E^D and reasoning parameters supplied by the master expert E^M. In this paper the following notation is introduced:

$$E^R_{\varsigma_i}\{[\alpha \ \equiv x] \Rightarrow c\} \mapsto E^B$$

that means the Reasoning-Expert $E^R_{\varsigma_i}$, based on expertise α supplied by the decision making expert, formulates a protocol c and invokes another expert. The symbol $\mapsto$ denotes that the new expert is invoked.

Let's assume that the decision is to be made about the reasoning stage "convex/concave/thin object". The Reasoning-Expert E^R (*ReasoningExp*) is invoked by the Master-Expert E^M. At the reasoning stage *ConvexConcaveThin* the Reasoning-Expert E^R makes a decision *DECISIONConvexConcaveThin()* by creating the Decision-Making-Expert E^D (*DMCvCoTh*) and invoking the method *MakeDECISION()*. Depending on the decision made by the decision-making expert one of the following stages of reasoning is selected: *s_CONVEX*, *s_CONCAVE*, or *s_THIN*. At each stage of reasoning the processing expert is invoked (*pMenagerExp->ProcesCONVEX()*, *pMenagerExp->ProcesTHIN()*) and decision is made about the further way of reasoning. The following code fragment shows how reasoning expert made decision at the reasoning stage *ConvexConcaveThin*.

```
void CReasoningExp::DECISIONConvexConcaveThin()
{
pDMCvCoTh=new CDMCvCoTh;
pDMCvCoTh->MakeDECISION(m_ExpertParam,
m_NumberDistanceRozklad);
switch (pDMCvCoTh->m_Type)
{
case s_CONVEX:
pMenagerExp->ProcesCONVEX();
DECISIONPolygonCurve();
m_EndAnalysisResult=pMenagerExp->m_SymbolResult;
break;
case s_CONCAVE:
pMenagerExp->ProcesCONVEX();
DECISIONPolygonCurve();
if(pMenagerExp->m_NumberConvexityPoints==0)
{m_EndAnalysisResult=pMenagerExp->m_SymbolResult; }
else
{m_EndAnalysisResult=pMenagerExp->m_SymbolResult;
DECISIONConcave(); }
break;
case s_THIN:
pMenagerExp->ProcesTHIN();
if(m_NumberPoints==2)
{DECISIONThin(); }
if(m_NumberPoints==4)
{pAAADMThinStar=new CAAADMThinStar;
pAAADMThinStar->MakeDECISION(m_NCPt, m_NNCPt); }
if(m_NumberPoints>4)
{pAAADMThinStarCztery=new CAAADMThinStarCztery;
pAAADMThinStarCztery->MakeDECISION(m_NCPt, m_NNCPt); }
}
```

Data Acquisition Expert (Manage-Expert)

The Manager-Expert E^Q receives messages from the Reasoning-Expert, selects and creates an appropriate Processing-Expert, stores data, deletes the

Processing-Expert, sends messages to the Reasoning-Expert to find how to carry out further processing and sends the acquired descriptors to the Reasoning-Expert.

During the reasoning process the new data needed at a given stage of reasoning are acquired. The Data-Acquisition-Expert (Manager-Expert) E^Q invokes an appropriate Processing-Expert E^P that is responsible for processing at a given stage of reasoning $E^Q\{[c_Q \equiv z_i] \Rightarrow c_P\} \mapsto E^P$ that means the Manager-Expert E^Q, based on expertise c_Q supplied by the Decision-Making-Expert, formulates a protocol c_P and invokes another Processing-Expert E^P. The Processing-Expert implements one of the processing methods Φ in order to obtain the data and a set of descriptors at a given stage of reasoning. The Processing-Expert E^P which implements the processing method Φ is denoted by one of the names of the possible shape classes e.g. E^P_Γ, E^P_Λ, E^P_Q, E^P_Θ, E^P_L, E^P_M or E^P_K. Based on the protocol received from the Manager-Expert the Processing-Expert selects an appropriate Algorithm-Expert.

The Processing-Expert invokes an Algorithm-Expert E^A which implements one of the image transformations Θ in order to obtain the required data. This process is described as follows:

$$E^P_{\varsigma_i}\{[c_P \equiv v_i] \Rightarrow c_A\} \mapsto E^A_k$$
$$E^A_k\{[w = \Theta_j]\} \Rightarrow \Pi$$

where Π denotes the data obtained by applying the image transformation Θ_j.

During reasoning process, a perceived object is at first transformed into a set of critical points $\amalg$ and next into the symbolic name η. Perceiving an object can be seen as a process of acquiring a new data and new skills. In order to fulfill the required task of acquiring the data and processing it in order to obtain a set of descriptors $\Im$, processing methods Φ are used. The processing method applies an image transformation Θ in order to transform the data into one of the data types. The image transformation Θ is the mapping from the one set called the domain of mapping into another called the set of mapping values. As a result of applying the image transformation into a set of critical points $\amalg$, a new set of critical points $\amalg'$, a set of transform numbers Δ or a set of mapping numbers Σ, is obtained. The descriptor transformation $\aleph$ is applied to find a set of descriptors $\iota \in \Im$ used to assign the perceived object to one of the possible classes Ω^η.

A reasoning process that is part of a visual reasoning process is performed passing the consecutive stages of reasoning. During each stage of reasoning a sequence of image transformations is applied in order to find a set of descriptors. The sequence of image transformations $\Theta_{\text{Ш}}^{\lambda} : \text{Ш} -> \text{Ш}$ that are used in reasoning process can be written as: $\lambda_{\alpha_1} : \text{Ш}^{\alpha_0} \to \text{Ш}^{\alpha_1}$ $\lambda_{\alpha_2} : \text{Ш}^{\alpha_1} \to \text{Ш}^{\alpha_2}$,...., $\lambda_{\alpha_M} : \text{Ш}^{\alpha_{M-1}} \to \text{Ш}^{\alpha_M}$ or as a composite given as $\Theta_{\alpha_1} \bullet \Theta_{\alpha_2} \bullet ... \bullet \Theta_{\alpha_M} : \text{Ш}^{\alpha_0} \to \text{Ш}^{\alpha_M}$, where Θ_{α_1} denotes one of the image transformations and $\bullet$ denotes the sequential operator. The reasoning involves processing by applying one of the image transformations, computation of the descriptors using a descriptor transformation and assigning an object to one of the possible classes. Although it was assumed that a visual object is represented by a binary image it is not the cause of a serious limitation to the presented method. The visual object that consists of parts of different colours is assigned into one of the colour classes and during processing stages these parts are interpreted as the new visual objects.

The reasoning involves transformation of the description of an examined object s when passing stages $\varsigma_0 \to \varsigma_1 .... \to \varsigma_N$, where ς_0 is the beginning stage, ς_N is the final stage of the reasoning process and $\to$ denotes the move to the next stage of reasoning. If at a given stage of analysis there is a need to acquire new data or make a decision about further processing an appropriate expert is invoked. At each stage of the reasoning ζ_i the following operations are performed:

- the processing transformation transforms the set of critical points

$\lambda_2 : \text{Ш}^1 \to \text{Ш}^2$,

- the descriptor transformation computes descriptors

$\iota_2 = \aleph_2(\text{Ш}^2)$,

- an examined object s is assigned to one of the possible classes

$[\iota_2 < T_2] \Rightarrow s \succ \Omega[\varsigma_2]$.

An example of assigning an object to one of the concave polygon classes is given in Fig. 8.2. The reasoning involves processing by applying one of the image transformations, computation of the descriptors using a descriptor transformation and assigning the object to one of the possible classes given by the description ς_i. The examined object given in Fig. 8.2 is assigned to the concave polygon class $Q^1[L^4](L^3)$ passing the following stages of reasoning: $Q \Rightarrow Q^m \Rightarrow Q^m[L^n] \Rightarrow Q^m[L^n](m \bullet L^i)$.

The stage of reasoning $\zeta_0 \equiv Q$:

- the processing transformation:

$$\lambda_B : \amalg^F \to \amalg^B \qquad \lambda_\aleph : \amalg^B \to \amalg^\aleph \; \partial_H : \amalg^\aleph \to \cap^H \; \partial_\aleph : \cap^H \to \cap^\aleph,$$

- the descriptor transformation: $\iota_C = \aleph_C(\cap^\aleph) = \frac{|\cap^\aleph|}{|\amalg^F|} = \frac{8}{59} = 0.14$,

- the rule: $[\iota_C > T_C] \stackrel{\sim}{\Rightarrow} s \triangleright Q \quad [0.14 > T_{0.05}] \stackrel{\sim}{\Rightarrow} s \triangleright Q$.

The stage of reasoning $\zeta_1 \equiv Q^m$:

- the processing transformation:

$$\lambda_\Psi : \amalg^B \to \amalg^\Psi \qquad \lambda_\Phi : \amalg^B \to \amalg^\Phi$$

- the descriptor transformation: $\iota_\Phi = \aleph_\Phi(\amalg^\Phi) = 1$

- the rule: $[m = \iota_\Phi] \Rightarrow s \triangleright Q^m \quad [m = 1] \Rightarrow s \triangleright Q^1$

The stage of reasoning $\zeta_2 \equiv Q^m[L^n]$:

- the descriptor transformation: $\iota_\aleph = \aleph_\aleph(\amalg^\aleph) = 4$

- the rule: $[n = \iota_\aleph] \Rightarrow s \triangleright Q^m[L^n] \qquad [n = 4] \Rightarrow s \triangleright Q^1[L^4]$.

The stage of reasoning $\zeta_3 \equiv Q^m[L^n](n \bullet L^{h^k})$:

the processing transformations:

$$\lambda_\Sigma : \amalg^B \to \amalg^\Sigma \; \lambda_{\Sigma_n} : \amalg^B -> \amalg^{\Sigma_k^i} \quad \lambda_o : \amalg^{\Sigma_k^i} -> \amalg^o$$

the descriptor transformation: $\iota_{\Psi^k}^k = \aleph_\Psi(\amalg^{\Psi^k}) = 3$

the rule: $[h^k = \iota_{\Psi^k}^k] \Rightarrow s \triangleright Q^m[L^n](n \bullet L^{h^k})$

$[h^1 = 3] \Rightarrow s \triangleright Q^1[L^4](L^3)$.

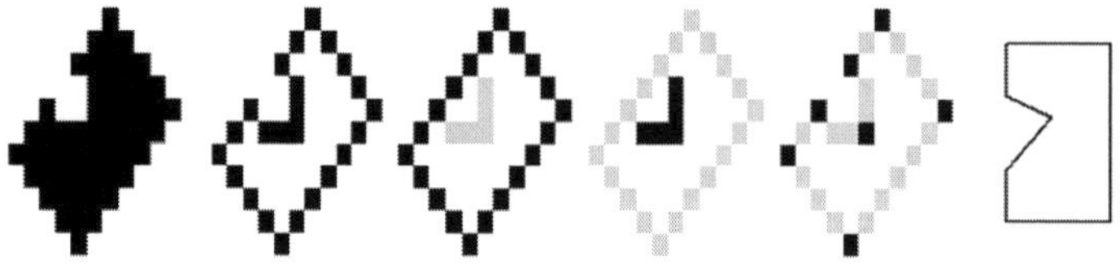

Fig. 8.2. Example of processing of an object: 1) an examined object given as a set of critical points $\coprod^F$, 2) the image transformation $\lambda_B : \coprod^F \to \coprod^B$, 3) the image transformation $\lambda_\aleph \circ \partial_H : \coprod^B \to \cap^H$, 4) the image transformation $\lambda_\Psi : \coprod^B \to \coprod^\Psi$, 5) the image transformation $\lambda_\Sigma \circ \lambda_{\Sigma_R} \circ \lambda_O : \coprod^B -> \coprod^O$, 6) an archetype of the class $Q^1[L^4](L^3)$ to which an examined object is assigned

For the description of the classes $Q^1[L^4](L^3)$ see [1], [125]. In the case of non-regular classes the reasoning process is more complex and is not presented in the book.

At the stage of reasoning $\zeta_0 \equiv Q$ the set of processing transformations was as follows:

$$\lambda_B : \coprod^F \to \coprod^B \qquad \lambda_\aleph : \coprod^B \to \coprod^\aleph \quad \partial_H : \coprod^\aleph \to \cap^H \; \partial_\aleph : \cap^H \to \cap^\aleph ,$$

Following notation that was introduced in [1], [126], the reasoning process can be expressed as a sequence of invoked experts. The stage of reasoning $\zeta_0 \equiv Q$ shown as a sequence of invoked experts is as follows:

$$E^R_\Lambda\{\} \mapsto E^R_Q,\; E^Q_\Lambda\{\} \mapsto E^P_Q,\; E^P_Q\{\coprod^F\} \mapsto E^A_F \mapsto E^A_B \mapsto E^A_\kappa \mapsto E^A_H \mapsto E^A_{\bar{\kappa}}\{\cap^\aleph\},$$
$$E^R_\Lambda\{\} \mapsto E^D_\lambda(Q),$$

that could be simplified by substituting invoked experts by a sequence of the invoked experts of the next processing action e.g. E^P_Q in $E^Q_\Lambda\{\} \mapsto E^P_Q$ is substituted by $E^P_Q\{\coprod^F\} \mapsto E^A_F \mapsto E^A_B \mapsto E^A_\kappa \mapsto E^A_H \mapsto E^A_{\bar{\kappa}}\{\cap^\aleph\}$. As a result, the first stage of reasoning can be represented in the following form: $E^R_\Lambda\{\} \mapsto (E^Q_\Lambda\{\} \mapsto E^P_Q\{\coprod^F\} \mapsto E^A_F \mapsto E^A_B \mapsto E^A_\kappa \mapsto E^A_H \mapsto E^A_{\bar{\kappa}}\{\cap^\aleph\}) \mapsto E^D_\lambda(Q) \mapsto E^R_Q$.

Similarly, all reasoning processes can be expressed as a sequence of invoked experts. This new method of the description of the performance of the system makes it possible to describe the performance of the complex system in the unified notation of the invoked experts. This description will be further elaborated to supply tools for testing the complex systems such as SUS.

The Manager-Expert E^Q, implemented as a *CMenagerExp*, receives messages from the Reasoning-Expert, selects and creates the appropriate Processing-Expert, stores data, deletes the Processing-Expert, sends messages to the Reasoning-Expert to find how to carry out further processing and sends the acquired descriptors to the Reasoning-Expert. The following code fragments show how *CMenagerExp* receives messages from the Reasoning-Expert, selects and creates an appropriate Processing-Expert, and stores data.

```
void CMenagerExp::FirstStageReasoning(POINT *Data,int NumberData)
{
MakeImagePoints(Data,NumberData);
MakeCenter(m_ImagePoints,m_NumberImagePoints);
ProcesPROLOG();
}

void CMenagerExp::MakeImagePoints(POINT *Data,int NumberData)
{
m_NumberImagePoints=NumberData;
m_ImagePoints=new POINT[m_NumberImagePoints];
for(i=0;i<m_NumberImagePoints;i++)
m_ImagePoints =Data[i];
}

void CMenagerExp::MakeCenter(POINT *Data,int NumberData)
{
m10=0; m01=0; m00= NumberData;
for(i=0;i<NumberData;i++)
{ x=Data[i].x; y=Data[i].y; m10=m10+x; m01=m01+y; }
XCenter=m10/m00; YCenter=m01/m00; m_Center=new POINT;
m_Center->x=(int)XCenter; m_Center->y=(int)YCenter;
}
```

The Manager-Expert *CMenagerExp::FirstStageReasoning*, after preliminary pre-processing of data (*MakeImagePoints* and *MakeCenter*) invokes the Processing-Expert *ProcesPROLOG()*. The method *MakeImagePoints* is invoked to store the critical points $\coprod^F$ that are obtained in the process of data acquisition of SUS. The critical points are stored as an array of *m_NumberImagePoints* points called *m_ImagePoints*. The method *MakeCenter* computes the centre of the examined shape that is stored as *m_Center*. The method *ProcesPROLOG()* that invokes the Processing-Expert is the method of the first stage of the reasoning process. At this stage the Processing-Expert called *ProcesProlog* is created and after processing by invoking *pProcesProlog->Update() ProcesProlog* this expert is destroyed. The methods *UpdateCurve(), UpdateConvex(), UpdateConcave(), UpdateParametry(), and UpdateDistan()* are applied to store the results of the processing process.

```
void CMenagerExp::ProcesPROLOG()
{
pProcesProlog=new ProcesProlog;
pProcesProlog->Update(m_ImagePoints,m_NumberImagePoints);
UpdateCurve();
UpdateConvex();
UpdateConcave();
UpdateParametry();
UpdateDistan();
delete pProcesProlog;
}
```

The Processing-Expert called ProcesProlog invokes algorithm experts E^A, AL-Border, ALConvex, ALConvexity, ALParametry, ALDistan. The Algorithm-Expert E^A implements one of the image transformations Θ in order to obtain the required data or descriptor transformations $\aleph$ in order to obtain a set of descriptors. The following code shows the processing by invoking Algorithm-Experts.

```
void ProcesProlog::Update(POINT *ImagP,int NoImagP)
{
pALBorder=new CALBorder;
m_pALConvex=new CALConvex;
pALConvexity=new CALConvexity;
m_pALParametry=new CALParametry;
m_pALDistan=new CALDistan;
pALBorder->Update(BorderPar, ImagP,int NoImagP);
m_pALConvex->Update(ConvexPar,pALBorder->m_CurveP,pALBorder-
>m_NoCurveP);
pALConvexity->Update(ConvexityPar,pALBorder->m_CurveP,pALBorder-
>m_NorCurveP, ImagP,NoImagP);
m_pALParametry->Update(ParamPar,ImagP,NoImageP,pALBorder-
>m_CurveP,m_NoCurveP);
m_pALDistan->Update(Distan,Par,ImageP,NoImageP);
deleteAll();
}
```

At the second stage of processing called ProcesHole(), the examined object is checked if it does not contain holes (belong to the cyclic class). In the case when the hole is found the complex iterative process is started. The Decision-Making-Expert MakeDecisionR makes a decision based on a set of rules and governs the reasoning process at the first stage of reasoning. The following code shows example of processing to find if an examined object contains the holes (if it is the member of the cyclic class).

```
void CMenagerExp::ProcesHole()
{
pProcesHole=new ProcesHole;
pMakeDecisionR=new MakeDecisionR;
pProcesHole-
>Up-
date(m_CurvePoints,m_NoCurvePoints,m_ImagePoints,m_NumberImagePoints);
m_NumberHoles=pProcesHole->pALHole->m_NumberOutPoint;
pMakeDecisionR->Update(m_NumberHoles);
if(pMakeDecisionR->Clas=CYCLIC)
ProcessCyclic();
m_HolesPoints=UpdatePoints(m_NoHoles,pProcesHole->pALHole-
>m_OutPoint);
delete pProcesHole;
delete pMakeDecisionR;
}
```

The Algorithm-Expert E^A implements the image transformations Θ in order to obtain the required data. For example, the image transformations $\Theta^{\alpha}_{\beta}[Ang]$ is represented by a specific method of computing the data Angle(), Angle_I(), Angle_II(), or Angle_III(). The image transformations $\Theta^{\alpha}_{\beta}[Ang]$ are specified by two parameters: α that denotes the type of data and β that denotes a processing algorithm. The specific method of computing of the data is selected by specifying parameters α and β implemented as a structure m_Parameters.

The important factor that characterizes the skills of the system is the ability to store, generate and process the images. Factors such as memory, speed and efficiency of the algorithms determine the basic visual skills of SUS. The Image-Data-Expert implements basic algorithms that read the visual object. The visual object is perceived as the image 1024x1024 pixels 24-bit colures and is transformed into 1024x1024 pixels (16-Color) for a silhouette, a line-drawing, a colors object and 1024x1024 pixels (256-Color) for a shaded-object. The exemplars are stored as images (1024x1024) (black and white). The exemplar e as one of the regions of a binary image is regarded as a set of pixels on the discrete grid (i,j). The exemplar is represented by a set of critical points $\amalg^F$ and implemented as an array of N points. An exemplar e of the class Ω is the binary realization of an archetype in the discrete space. The shape classes are given by the symbolic names, for example, the symbolic name of the object  is given as $A[L_E^3](3L_O^3)$. This symbolic name η_k is easy to transform into the symbolic name (SUS notation) given as A3_L3_AE_L3_O_L3_O_L3_O. The symbolic names (in SUS notation) are stored in the text format and are processed by the Symbolic-Name-Expert. The Symbolic-Name-Expert reads the symbolic names from the text file and stores these symbolic names as an array of strings.

The symbolic names are used to generate the exemplar e from the selected class (given by the selected symbolic name) by invoking the Exemplar-Generating-Eexpert. The Exemplar-Generating-Expert generates exemplars e that belong to one of the shape classes based on the template Λ^n given in the form of the generic vertices $\Lambda^n = [(x_1, y_1),...,(x_N, y_N)]$. These exemplars are used during learning in a self-correcting process [127]. During the self-correcting process the Self-Correcting-Expert makes decision if the generated object $o_j \mid \omega_k \in \Omega[\omega_k]$ is to be assigned to the class $\Omega[\omega_k]$. Exemplars from the selected class are also generated as part of the response to the command such as "draw pentagon", "draw concave figure". The Exemplar-Generating-Expert is invoked when an exemplar of the class $\Omega[\omega_k]$ is needed. The Exemplar-Generating-Expert invokes an appropriate Model-Expert to specify parameters of the model. Specification of the parameters imposses constraints on the values of the attributes of the model and defines an archetype of the class $\omega_k \in \Omega^k$. The model expert defines the archetype of the class in the form of templates. During the process of exemplars generation the Model-Expert invokes an Algorithm-Expert that applies the different methods to generate the required exemplar. For example, the command "draw pentagon" invokes the Algorithm-Expert with a proper method to generate an exemplar of the pentagon category. Without further specification the generated exemplar can be any pentagon. The exemplar from the class such as the convex polygon class L^n can be relatively easy to generate based on the specifications of the selected parameters. The name of the class such as the convex polygon class, denoted as L^n, is called the symbolic name η. Learning of the complex object requires using many parameters to describe the specific class. This complex description is represented by a symbolic description κ. During reasoning process, the symbolic name η is extracted from the symbolic description κ. The symbolic description κ is an intermediate form that has many additional specific data about the perceived object. The symbolic description κ is used to reason about the specific ontological categories to which the examined object can belong. For example, the object O1  stored in the file S1.bmp as a bitmap (1024x1024) is transformed into a symbolic description in the form of the string:

"[A3][[|L3|AE|]|S79||B100,99,99||A60,61,60||G248||@2691|]]{[|L3|O|]|S52||B58,100,57||A29,30,120||G76||@395|]}{[|L3|O|]|S52||B57,100,58||A30,29,120||G76||@396|]}{[|L3|O|]|S53||B100,58,57||A29,120,30||G76||@417|]}".

The symbolic description κ as well as the symbolic name η (SUS notation) are stored in the text format and are processed by the Symbolic-Name-Expert. The

Symbolic-Name-Expert reads the symbolic names from the text file and stores these symbolic names as an array of strings.

As it was described, the visual object (phantom) u_i is perceived as an image 1024x1024 pixels 24-bit colures and is transformed into 1024x1024 pixels (16-Color) for the category of a silhouette, a line-drawing, a colors-object and 1024x1024 pixels (256-Color) for the category of a shaded-object. It is assumed that the ontological categories such as the figure or letter categories are given by objects that are members of the silhouette category. Also some objects from the real world object category such as the tools category can be given by objects that are members of the silhouette category.

Understanding of the visual object from one of the ontological categories requires learning of the visual concept of this category. The ontological category v_i is represented by a set of visual objects $O^i[v_i]=\{o_1, o_2, ..., o_n\}$. The visual knowledge of the category v_i is learned as a visual concept of this category and is stored as a set of symbolic names $\varphi_c=\{\eta_1, \eta_2, ..., \eta_n\}$. It is assumed that a set $O^i[v_i]$ represents all visual aspects of the category v_i. During learning stage, at first, the representative sample of objects from the category v_i is selected, then for each object the symbolic name η_i is obtained and finally a visual concept of this category as a set of symbolic names $\varphi_c^j=\{\eta_1, \eta_2, ..., \eta_n\}$ is obtained. The visual concept for each learned category is stored as an array of symbolic names $\varphi_c^1, ..., \varphi_c^K$, where K is a number of learned categories.

During understanding of the perceived object, an examined object u is transformed into the symbolic name η and next a set of similar objects is searched to find the symbolic name of the category that was learned previously. When the symbolic names are found, all visual concepts $\varphi_c^1, ..., \varphi_c^K$ that have as a member this symbolic name are extracted. In the case when K=1, the reasoning is stopped and the name of the category v_i is used as the name of the examined object u.

All learned, for this category v_i, the non-visual knowledge is now accessible and can be used in the thinking/understanding process. In the case when K=0, there is a need to use the generalization process to find the most similar objects. In the case when K>0 there is a need to use learned the non-visual knowledge to properly classify the object to one of the object categories. The most important part of the visual understanding is the visual reasoning that utilizes the different image processing methods. These methods, implemented in the form of the image processing algorithms, determine the complex visual skills of the SUS.

8.1.2 Implementation of the New Skills –The Category of Visual Objects

In this Section learning of the knowledge of the visual object is presented. As it was described, the knowledge implementation is concerned with learning of the visual and the non-visual knowledge from the different categories of object. Learning of the knowledge of the visual object can be seen as learning of the specific perceptual skills to acquire data from an image. During learning process not only the new data are gathered and used in learning of the new knowledge but also the new shape classes are established, the new image transformations and the new reasoning experts are implemented. The visual knowledge is learned in the form of the symbolic representations of the visual aspects of the categories of the visual objects. The ontological category v_i is given by the name n_i and is represented by a set of visual objects $O^i[v_i] = \{o_1, o_2, ..., o_n\}$. A visual object (phantom) u_i is perceived as the image 1024x1024 pixels 24-bit colures and is transformed into the image 1024x1024 pixels (16-Color) for a silhouette, a line-drawing, a colors object and the image 1024x1024 pixels (256-Color) for a shaded-object. As it was described, the ontological categories such as the figures or letters categories are represented by objects that are members of the silhouette category.

The symbolic description of the visual aspects of categories of the visual objects is represented by the symbolic names. During learning, the symbolic name of shape classes is transformed into the SUS-symbolic name represented in the form of strings (SUS notation). In some cases the symbolic names of shape classes can be very easily translated into the SUS symbolic names. However, due to the perceptual error connected with the perceptual ability of SUS, there is a need to test the learned visual knowledge for cases that can produce an error. For example, the object  has the symbolic name $A[Q^2[L_R^4](2L_R^4)](L_R^4)$ that is translated into the string A1_Q2_u_u_L4_R_q1q1_L4_R_L4_Xn during the learning process. The object  has the symbolic name $Q^2[L_R^4](2L_R^4)]$ that is translated into a string Q2_u_u_L4_R_q1q1_L4_R during the learning process. However, for objects  and  the symbolic name needs to be learned and tested because an object from the curvilinear class can have the different generic polygons and their symbolic names can differ from the symbolic name obtained in the form of strings (SUS notation). The SUS-symbolic name obtained during the learning process for the object given by the symbolic name $Q^2[M^1[L_R^4]](2L_R^4)]$  is as follows: Q2_u_u_M1_L4_WWn_1q1q_000k_L4_R_L4_XnA3a_r2E_r2e_g2E_g2e; and for the object given by the symbolic name $Q^2[M^1[L_R^4]](2Q^1[M^1[L_M^4]]M_i^1)]$  is "Q2_Q1_u_Q2_u_u_M1_L4_Ha_1q1q_000k_L4_Ha_aqqq_M1_L4_Wgn_aqaq_M 1_M1/Lt". As we can see there is a difference in symbolic names of shape classes and symbolic names in the form of the string.

Learning of the knowledge of the objects from the different structural categories such as the element category, the pattern category or the picture category require application of the different skills and knowledge for processing, visual reasoning and visual understanding. The members of pattern category derived from the category of visual symbols such as the category of mathematical expressions, the category of texts, the category of musical texts or the category of engineering schemas need to be interpreted by using complex methods that involve huge amount of non-visual knowledge. A member of the element category such as the minerals category requires the complex sensory data in order to recognize and understand the mineral object.

Learning of the knowledge of the selected category of the visual objects requires learning both visual and non-visual knowledge. The visual knowledge can be learned from perceptual categories such as a silhouette, a line-drawing, a color-object or a shaded-object. In present research, a silhouette category is used for learning of the visual concepts of the selected categories of the visual objects such as the sign category. The reason for using the silhouette category is that most objects from the sign category belong to the silhouette category. Also at the first stage of the categorical learning [128], relatively simple objects from the category of real world objects are learned. Objects from the line drawing, colored objects and shaded objects categories are transformed into the silhouette category by applying the proper image transformation. For learning of the knowledge of ontological categories the sign category was selected. The reason for the selection of the sign category is that most objects from the sign category are "visually" well defined and usually belong to the silhouette category. The knife category that is derived from the tool category is selected to learn knowledge of parts decomposition. Objects from the knife category are relatively simple and have usually well visible parts.

The different categorical learning methods described in Section 7.1.3 such as the learning by the small alternation of the visual objects (LSA), learning from the simple to complex (LSC), learning by simplification of the complex objects (LSCO), learning from parts (LP), and learning parts decomposition (LPD), were tested independently and next these methods were applied for learning of the selected ontological categories.

For learning of the selected ontological categories v_i a set of visual objects that are representatives of this category is selected. Each 3D object is transformed into the phantom that is the 2D representation (e.g photograph) of the object and next into the image 1024x1024 pixels (16-Color). The visual knowledge of the category v_i is learned as the visual concepts of this category and is stored as the part of the Visual-Concept-Expert. The visual concept is a set of symbolic names that are extracted from the symbolic descriptions. The symbolic description is an intermediate symbolic form that has many additional specific data about the perceived phantom. The symbolic description is used to reason about the specific categories to which an object can belong.

Learned symbolic descriptions and symbolic names are stored in the form of text files as the part of the Visual-Concept-Expert. The symbolic name is used to describe an object and to make inference about the object at an intermediate level of understanding. At the intermediate level of understanding the object of a given category is interpreted in terms of concavities and holes and is represented by the intermediate description. The intermediate description of a given category is stored as part of the knowledge of the Intermediate-Level-Name-Expert. At the intermediate level of understanding an object is also described in terms of the structural archetype and can be decomposed into parts by applying knowledge of parts decomposition. In order to decompose the object into parts correctly, the proper object decomposition transformations are implemented as the Part-Decomposition-Expert.

The important part of learning of the non-visual knowledge is to learn naming of an object that assigns name of the ontological category to the visual object. Learning of the name of the object from the selected category is connected with learning of the knowledge of the categorical chain and learning of the knowledge of the knowledge schema. The categorical chain is implemented as an array of interconnected names of the categories. During learning at the first stages of learning there is no need to supply names for all learned categories at all categorical levels. During incremental categorical learning names of categories that were not learned can be added and new inter-categorical links can be established.

Learning by the small alternation of the visual object

In this Section, the ability of SUS of learning by applying the small alternation method (LSA) is described. Learning by the small alternation (LSA) is to select one visual object (basic object) and next, by changing its visual features, learning of the visual knowledge from a sequence of generated objects. At first, the basic visual objects were selected and next alternations were made in order to obtain visually more complex objects. Figs. 8.3, 8.4 and 8.5 show examples of the visual objects used for learning by the small alternation of the objects. For testing LSA method, the following four objects were used (letters ⊔ and T , and two figures triangle and circle). Figs. 8.3 and 8.4 show the objects {letters: ⊔, T } used for learning and also their alternated versions. Examples shown in Figs. 8.3 and 8.4 were used for learning of the category of letters in the context of similar objects. At first, the most typical objects from the letter category were selected and next, by alternating some visual features, the objects that can be members of this letter category or other ontological categories were learned. Fig. 7.10 (given in Chapter 7) shows the object △ that was selected for performing the alternations and the way in which alternation process was performed. Fig. 8.5 shows the object ○ that was selected for an experiment and the alternations performed that lead to obtaining the objects from the sign category.

Fig. 8.3. Examples of objects generated during learning by alternating the selected object: the letter U

Fig. 8.4. Examples of objects generated during learning by alternating the selected object: the letter T

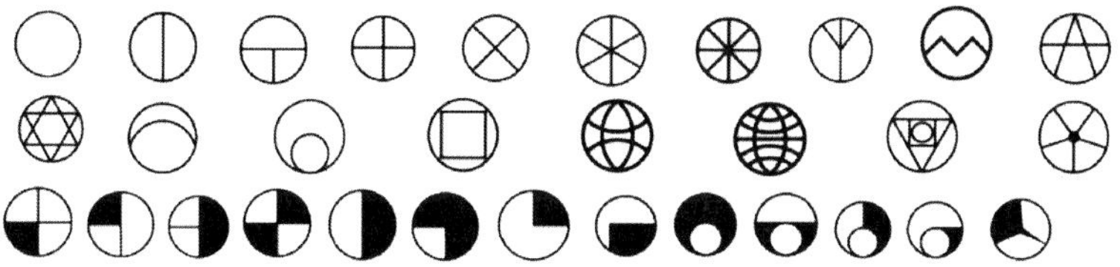

Fig. 8.5. Examples of objects generated during learning by alternating the selected object:

Learning From the Simple to Complex

Learning from the simple to complex can be regarded as a special case of learning by the small alternations of the visual object. The complex object that is learned is decomposed into simple parts as shown in Fig. 7.11 (given in Chapter 7). At first the symbolic names for the simple parts are obtained and next the more complex objects are learned. During learning process the new classes are established and new image transformations are added. Learning from the simple to complex is applied during learning of the sign category and the knife category. Fig.7.12 (given in Chapter 7) shows examples of objects from the knife category that were obtained as the result of simplification of the object and were used for learning of the knife category.

Learning by Simplification

Learning by simplification of the complex object (LSCO) is similar to learning from the simple to complex. The simplification of the complex object is based on an application of the image transformations such as the polygonization (removing curved segments), filling the holes or filling residuals (concavities). For testing purposes, ten cyclic objects were selected. The generalization transformations were applied (filling holes) and the symbolic names that were obtained were stored as linked arrays. Learning by simplification of the complex object is applied for learning of the sign category. Examples of objects used for testing the LSCO method have been shown in Figs. 7.13 and 7.14 (in Chapter 7). Examples of objects used for learning ontological categories are given in Section 8.1.2.1 that describes learning of the knowledge of the sign category.

Learning from Parts

Learning from parts (LP) is to combine the symbolic name of the learned object from the previously learned symbolic names of objects (parts). The result of learning of the visual object is the symbolic name that is stored as part of the visual concept. Symbolic names that were obtained during learning process are used to form the new symbolic name by transforming strings according to selected generalization rules. Learning from parts (LP) is used in order to facilitate learning of the big number of objects. Result of learning by applying the LP method can be seen as objects generated by some rules of generalization. For testing of the LP method the 10 objects O1 were combined with 5 objects O2 to form the new object O0. Examples of objects used in the testing process are shown in Figs. 7.15 and 7.16 (Chapter 7).

Learning Part Decomposition

During learning based on the part decomposition method (LPD) the image transformations are implemented to decompose an object into parts and next the symbolic names of parts that are learned are linked with the symbolic name of the object. During testing LPD method these objects that were used in the testing of the LP method, were also used in testing of the LPD method, Figs. 7.15 and 7.16 (Chapter 7). For the class of objects, that has common decomposition property, the image transformation is implemented and next the testing is carried out by comparison of symbolic names of parts after application of the implemented image transformations. Figs. 7.18 and 7.19 (in Chapter 7) show examples used in testing the LPD method. Application of the LPD method for learning of the knife category is described in section 8.1.2.5.

Learning of the Different Categories of Objects

In this book, only selected categories of objects used to learn knowledge of the different ontological categories, namely the signs category, the arrows category, the road signs category, the letters category and the knife category, are presented. The arrow category and the road signs category are specific categories derived from the signs category, but these categories are learned independently from other objects from the signs category. Important part of learning of the visual concepts of the ontological category is to select a prototype and all possible visual objects that are representatives of the prototype. At the first stage of learning the prototype can be represented by objects with the most characteristic visual features of the prototype. The result of learning of the prototype is a set of symbolic names called the visual concept of the prototype. The prototype selection is part of the general problem that is topic of research of prototype theory. The overview of the linguistic prototype theory can be found in [129].

8.1.2.1 Implementation of the New Skills – The Sign Category

In this Section, learning of the knowledge and skills of the specific categories derived from the sign category is presented. Knowledge of the specific categories derived from the sign category is learned at the prototype level. At the first stage of learning, the prototype is represented by only a few visual objects so the visual knowledge is covering only a part of the visual domain of the prototype. Learning of the knowledge of the category of signs involves learning both the visual knowledge and the non-visual knowledge. Learning of the non-visual knowledge is to learn a categorical chain as well as knowledge of a knowledge schema of the category of signs. Knowledge of the categorical chain is stored as linked arrays of names of categories of the categorical chain. The knowledge schema includes the name of the category and a meaning of the sign category. However, for some categories, the knowledge schema can be incomplete and can include only the name of the category. For example, for the category Ghanian adinkra symbol knowledge of the knowledge schema is stored in arrays name[], meaning[], as name[akobean] meaning[vigilance and wariness] or name[akokoan] meaning[mercy, nurturing]. For other categories, knowledge of the knowledge schema is stored as the name of the prototype in the array name[]. For example, the category of alchemistic symbol is given by its name: brass, silver, salt, water, honey, spirit of substance, glass, arsenic, four elements, lime, nitrogen, sulphur, mercury, brass, iron, antimony, orichalcum or dissolvent, Lye, decoction. During further stages of learning complete knowledge of the knowledge schema will be learned as the part of the knowledge of the Sign-Knowledge-Expert.

At the first stage of learning, the selection of categories of signs for learning was based on the criterion of availability of the representatives of these categories. Most of signs and symbols used to learn of the visual knowledge of the signs category were incorporated from existing literature [130], [131], [132], [133], [134], [135]. Visual objects were scanned or produced by drawing using TurboCAD Designer. The following specific categories were used to learn of the visual knowledge of the signs category: the category of Ghanian adinkra symbols, the category of alchemistic symbols, the category of Kabbalistic signs, the category of zodiac signs, the category of planet symbols, the category of gods signs, the category of rune symbols, the category of modern signs, the category of international weather symbol, the category of international weather symbol (cloud symbols), the category of international weather symbol (sky coverage), the category of astrological signs, the category of corporate emblems, the category of trademarks, the category of ideological symbols, the category of ancient symbols, the category of electronics symbols, the category of card suits, the category of the musical symbols, the category of currency symbols, the category of the special signs, and the category of mathematical symbols. Examples of visual representatives used for learning of the visual knowledge (visual concepts) of prototypes of different specific signs categories are shown in Figs. 8.6-8.22. It was assumed that each specific category (prototype) is represented by one or more than one visual object. In order to represent a prototype, up to five objects was selected. Each prototype from the specific signs category is given by its name that is stored as the part of the knowledge schema.

In the case when the different prototypes have the very different visual representatives, the learning is connected with learning of the visual concept and the visual schema. In the case when the different prototypes have similar visual representatives, specialization described in previous Sections was applied. SUS capability to learn knowledge of different categories was tested in the context of all knowledge of learned categories. In the case when the learned object referred to different ontological categories the proper link to the knowledge schema was established and specialization, if needed, was applied.

Fig. 8.6. The category of Ghanian adinkra symbol 1. aobean 2. akokoan 3. akoma ntoso 4. hwemudua 5. nkonsonnkonson 6. nyame birbi wo soro 7. nyame nnwu na mawu 8. nsorroma 9. ea 10. adinkrahene 11. sanfoka 12. denkyem 13. gye nyame

Fig. 8.7. The category of alchemistic symbol 1.brass 2. silver 3. salt 4. salt water 5. hney 6. spirit of substance 7. glass/arsenic 8. four elements 9. hour time 10. magnesium 11. lime 12. nitrogen 13. sulphur 14. mercury 15. brass 16. vitrol 17. iron 18. antimony 19. orichalcum 20. lye 21. decotion 22. crucible

Fig. 8.8. The category of kabbalistic sign 1. fire 2. Earth 3. air 4. algol 5. algorab 6. alphecca 7. amalgam 8. hagith 9. polaris 10. phalec 11. sirius 12. bethor

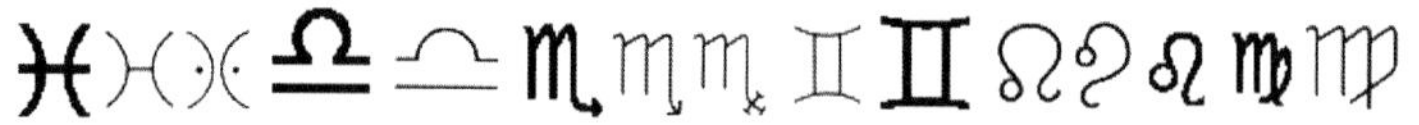

Fig. 8.9. The category of zodiak signs 1-3. pisces 4-5. libra 6-8. scorpio 9-10. Gemini 11-13. Leo 14-15. virgo

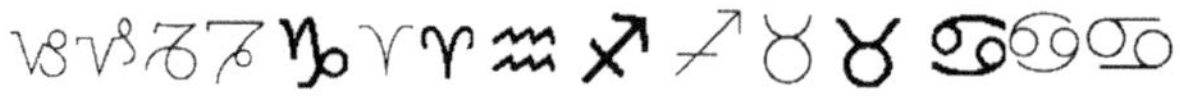

Fig. 8.10. The category of zodiak signs 1-5. capricon 6-7. aries 8. aquarius 9-10. sagittarius 11-12. taurus 13-15. cancer

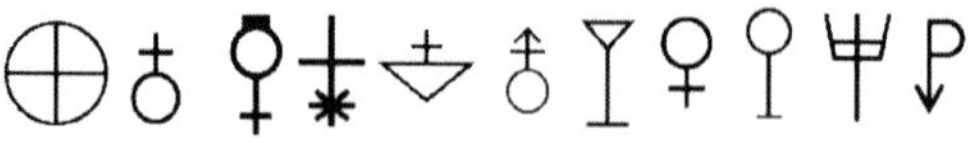

Fig. 8.11. The category of planet symbols: 1-2. earth 3-4. juno asteroid 5. pallas asteroid 6. mars 7-9. venus. The category of gods signs: 10. poseidon 11. hermes

Fig. 8.12. The category of rune symbols 1. madr 2. beorc 3. yew 4. thorn 5. wynn 6. daeg 7. day 8. deth

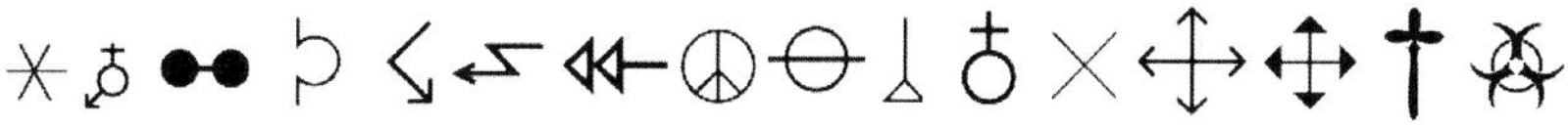

Fig. 8.13. The category of modern signs: 1. snow 2. hermaphrodite 3. eclipse of the moon 4. cave 5. lightning 6. thunder lightning 7. fast movement 8. antinuclear campaign emblem 9. diameter or average number 10. woman 11. chapel 12. multiplication, confrontation, annulment, cancellation, opposition, obstruction, mistake 13-14. cross arrowed 15. heraldic dagger 16. biohazard warning

Fig. 8.14. The category of international weather symbol (IWS) 1. lighting 2. haze 3. light intermittent snow. The category of international weather symbol IWS (clouds) 4. cumulus 5. cumulonimbus 6. stratocumulus 7. cirrus 8. allocumulus. The category of international weather symbol IWS (sky coverage) 9. cloudless sky 10. slightly covered sky 11. cloudy sky 12. very cloudy sky 13. overcast sky 14. obscured sky

Fig. 8.15. The category of astrological signs: 1. procyon 2. quintile 3. quincunx. The category of New Age symbols 4. crystal. The category of Masonic symbols 5. cube The category of ideological symbols 6. thunderbolt 7. swastika 8. hammer and sickle 9. antinuclear campaign emblem 10. anarchism 11. communism

Fig. 8.16. The category of corporate emblems 1. Mitsubishi 2. BMW 3. Chrysler 4. citroen 5. Mercedes

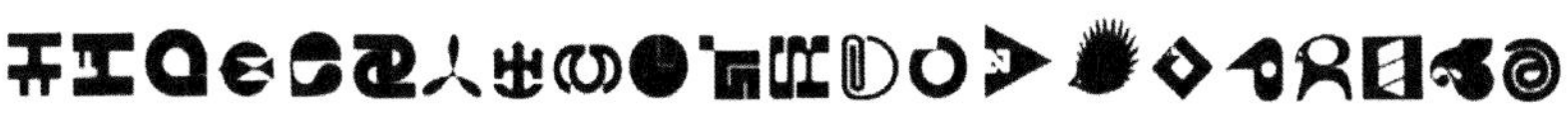

Fig. 8.17. The category of trademarks

Fig. 8.18. The category of ancient symbols 1. swastika 2. eye of fire 3. eye of dragon 4. empedocless elements, fire 5. hnefatal 6. nannan 7. circle with vertical line 8. circle intersected with horizontal line 9. circle on vertical line 10. swastika 11. Fire 12. dose of medicine 13. China symbol of how they viewed their civilization 14. hourglasys

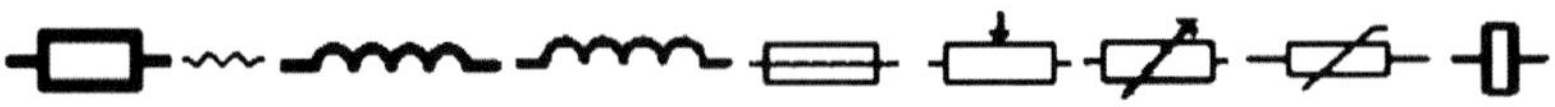

Fig. 8.19. Te category of electronics symbols 1-2. resistor 3-4. inductor 5. fuse 6-7 variable resistor 8. termistor 9 crystal oscilator

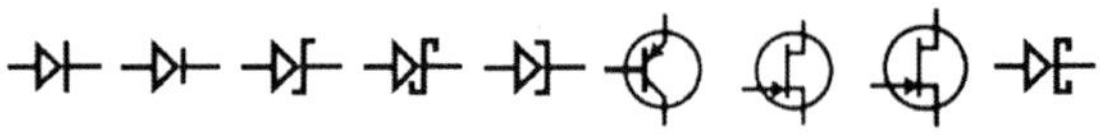

Fig. 8.20. Te category of electronics symbols 1-2 diode 3. Zener Diode 4. Schottky Diode 5. Tunnel Diode 6. PNP Bipolar Junction Transistor (BJT) 7-8 NPN Junction Field-Effect Transistor (JFET) 9. para-diode

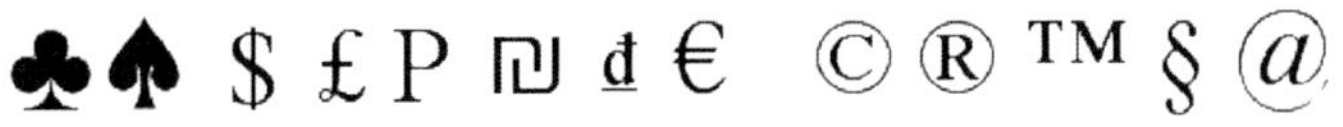

Fig. 8.21. The category of card suit, the category of currency symbols and the category of the special signs

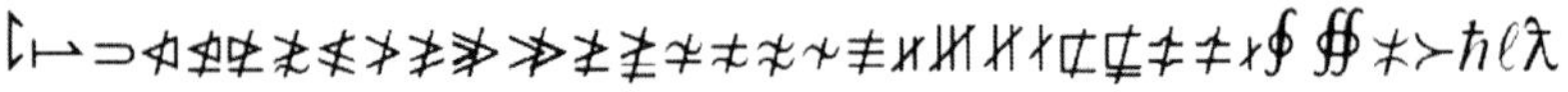

Fig. 8.22. The category of mathematical symbols

8.1.2.2 Implementation of the New Skills – The Letter Category

In this section learning of the knowledge of the letters category is presented. Learning of the knowledge of the letters category is to learn of the visual knowledge of letters from all existing writing systems and the non-visual knowledge connected with using a letter during reading and understanding of the text. At the first stage of learning of the letters category, categories of letters from existing writing systems were selected. Learning of the knowledge of the letters category depends on the way of a graphical designing of a letter as well as on the way of writing (hand writing). Due to many different font categories, learning of an object from a selected letter category requires learning of a huge number of visual objects. During learning of a letter category the selected categories of fonts from four alphabets Latin, Cyrillic, Hebrew and Greek were used. Meaning of a visual object from a letter category depends on an orientation of an object in the context of a "screen plane". For example an object Δ can have the different meaning depending on its orientation e.g, Δ means - Greek letter, ∇ means- mathematical symbol, ▷means - mathematical symbol. In order to learn these letters the new shape classes that capture all visual features connected with placement of the letter on the screen are established.

During learning of letters categories, the different techniques of categorical learning are utilized. The capital letters T and I of the Roman alphabet were selected to learn the prototypes of these letters. For each prototype one representative was selected, the visual concept was obtained and stored as the part of the knowledge scheme. Fig. 8.23 shows examples of letters used in learning of the prototypes of the letter fonts.

Fig. 8.23. Examples of members of the different categories of different fonts of the letter "T" and letter "I" used in learning of prototypes of the letter fonts

The categories of letters for which the representatives are "complex objects" are learned using the method of learning from the simple to complex (LSC). This method is based on learning by the alteration of features of the selected simple object. During learning of a category of letters by applying the LSC method, the important part of learning is naming process. Fig. 8.24 shows examples of objects used for learning a prototype O of a letter . Some objects that are similar to the object  can be called (O-like) but not by the name of this prototype (e.g. an object  is called -like). Some objects that are not similar to the learned prototype are named as (O-derived) (e.g.  is called -derived). During further learning the object  can be named by the name of the prototype from other ontological categories.

Fig. 8.24. Examples of objects used for learning prototype "O" of the letter 

For learning of the knowledge of the letter category the letters from the Roman alphabet, the Greek alphabet, the Cyrillic alphabet, the Arabic alphabet and the Hebrew alphabet were selected. During learning process that utilized the LSC method, the learned letter was "simplified" and each object that was learned was named "letter O", "O-like object", or "Q-derived object". For each prototype, a number of generated simplified objects were from 3 to 25 objects. For some learned objects, new specific shape classes were established and new image transformations were implemented. Learned objects were tested in the context of all learned visual objects from the different ontological categories. Figs. 8.25 and 8.26 show examples of objects from selected letters categories used to learn the knowledge of letters category.

Fig. 8.25. Example of objects used to learn the different letters by applying the LSC method

Fig. 8.26. Example of objects used for learning of the different letter categories

8.1.2.3 Implementation of the New Skills – The Arrow Category

In this Section learning of the knowledge of an arrow category is presented. An arrow category is a category derived from both the signs category and the real world objects category. In this book, the arrow category is regarded as the category derived from the signs category and refers to the visual objects which we usually call an arrow. The prototype is established at the general level of the arrow category. Knowledge of the arrow category is learned based on visual features of objects from the arrow category. It is assumed that all objects that share some visual features of the common arrow (member of the arrows category) are learned as the part of the visual concept of the arrows category. This type of learning make it possible to name an object that is similar to a common arrow by using the term "arrow-like". At first, the representatives of the arrows category (prototypes) are selected and arranged according to their visual complexity (120 visual objects). Next, for all visual objects the symbolic names are obtained. Figs. (8.27-8.30) show examples of visual objects used for learning of the visual concept of the arrow category.

During learning of the visual knowledge there is a need to learn objects that have very similar symbolic names but the different visual appearance. In order to cope with these differences there is a need to establish new specific classes. Figs. 8. 30 (b-l) show cases that have similar symbolic names but are different objects. Objects shown in Fig 8. 30 (b-l) are not members of the arrows category. Objects shown in Figs. 8. 30 (b-l) are differentiated by applying the different properties of the object using specialization and generalization. For example, the object in Fig. 8.30a is a member of the array category and is given by the symbolic name $Q^2[L^5][2L^3]$. Objects in Fig. 8.30 (h-j) are objects from the class $Q^2[L^4][2L^3]$ and objects in Fig. 8.30 (k,l) are objects from the class $Q^2[L^6][2L^3]$. These objects are not members of the arrows category, but can be named as "arrow-like" and can be easily distinguished based on the differences in symbolic names. Objects shown in Fig. 8.30 d-g have the symbolic name $Q^2[L^5][2L^3]$ but these objects are not similar to objects from the arrow category. Objects shown in Fig. 8.30 d,e can be distinguished from the object that have the same symbolic name (shown in Fig. 8.30 a) by using the side coding system (described in Chapter 7). Objects shown in Fig. 8.30 f,g can be distinguished by introducing of the new parameters of the new established specific class. Objects shown in Fig. 8.30 b,c have the symbolic name $Q^2[L^5][2L^3]$ that is the same as for the arrow shown in Fig. 8. 30 a. These objects can be distinguished by applying complex class notation. According to this notation the arrow in Fig. 8.30 a has the symbolic name $C[L^3, L_R^4]$, whereas objects in Fig. 8. 30 b,c have the symbolic name $C[L^3, L_M^4]$.

The following specific arrow categories are derived to learn of the visual knowledge of the arrows category: the category of arrows with straight lines - one head, the category of arrows with straight lines - more than one head, and the category of arrows with curved lines (see Figs. 8.27-8.29). Using objects that are members of these categories for learning of the knowledge of the arrow category make it possible to learn objects that are similar. The specific categories of arrows such as the arrow $\longrightarrow$, that is a member of the category of mathematical symbols, are learned in the context of learning of the knowledge of the category of mathematical symbols.

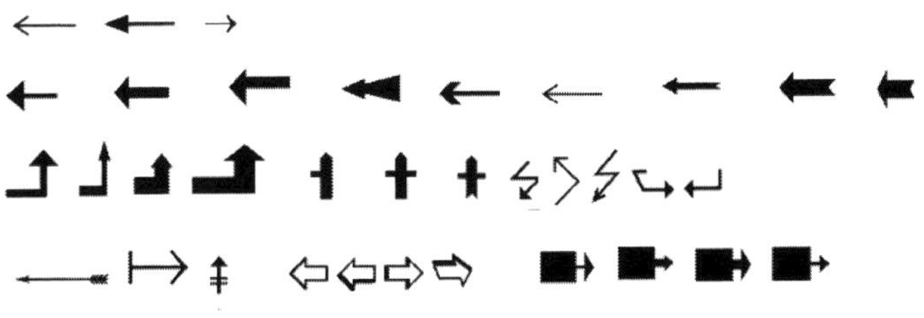

Fig. 8.27. The category of arrows with straight lines one head

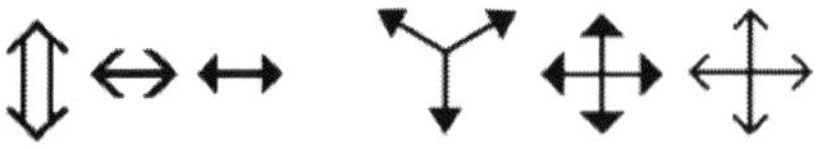

Fig. 8.28. The category of arrows with straight lines more than one head

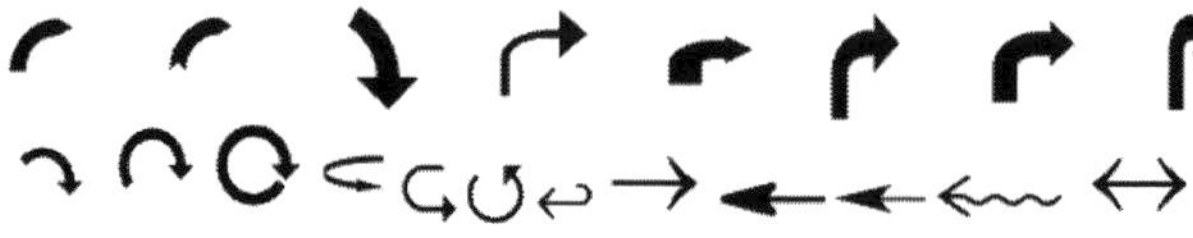

Fig. 8.29. The category of arrows with curved lines

Fig. 8.30. Object that has similar symbolic name but the different visual appearance

8.1.2.4 Implementation of the New Skills – The Road Signs Category

In this Section learning of the knowledge of the category of road signs is presented. The category of road signs is derived from the category of symbolic signs and the category of real world objects. The non-visual knowledge of members of the category of road signs is connected with their meaning and is learned as knowledge of a knowledge schema. The meaning of the road signs depends on the meaning of their parts. A member of the signs category consists of two parts: a background and a figure. Both parts of the object from the signs category, the figure and the background, have their specific meanings. Meaning of the background (e.g. circle, triangle) refers to the general category of the road signs. For example, an octagon is used for the category of stop signs and a triangle for the category of warning signs. The meaning of a figure is regarded in two aspects, the first aspect (the first meaning) and the second aspect (the second meaning). The first meaning refers to the visual object as a member of the category of real world objects or the category of letters. For example, the visual object  is interpreted as the member of the category of letters, a letter 'P', whereas the visual object  is interpreted as the member of the category of real world objects, a cup. The second meaning refers to the category of road signs. For example, a visual object  is interpreted as a member of the category of road signs; the cup indicates that there is a café near by. The learning of the signs category depends on knowledge learned in the context of other categories such as the category of letters or the category of real world objects. Knowledge of road signs used in understanding of the category of road signs is learned in the context of other categories such as the crosses category, the arrows category, the letters category, the vehicles category, the animals category, the men category or the tools category. The letter category is learned in the context of the word category. During learning of the signs, there is a need to learn knowledge of the words category and linguistic knowledge how to compose a word from the letters as well as the meaning of the word. Learning of the signs requires to learn the letters of the specific alphabet and to learn linguistic knowledge to understand the word from any existing language. Learning of the knowledge of the category of road signs requires learning of the knowledge of the category of ciphers and category of numbers, as well as learning of the knowledge of the category of the conventional representations of the animals. Fig. 8.31 shows examples of objects of the category of road signs.

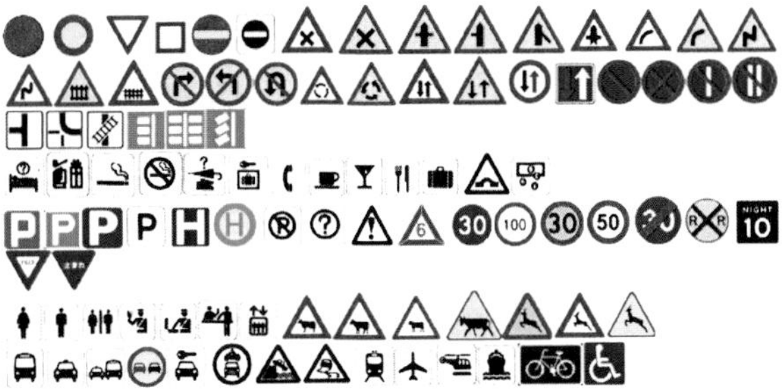

Fig. 8.31. Examples of objects of the category of road signs

Learning of the knowledge of the category of road signs is to learn of the visual knowledge as well as the non-visual knowledge related to this category. The non-visual knowledge is given as a knowledge schema of the specific category derived from the category of road signs. The specific eight categories of signs were derived based on annex 1 of the Vienna Convention on Road Signs and Signals of November 8, 1968. The specific categories of road signs are: 1. Danger warning signs (1DW) 2. Priority signs (P) 3. Prohibitory or "restrictive signs (R)" 4. Mandatory signs (M) 5. Special regulation signs (SR) 6. Information, "facilities", or "service" signs (I) 7. Direction, "position", or "indication" signs (D) 8. Additional panels (A). Learned knowledge of the knowledge schema is stored in the text file in the form of arrays name_roadSign[], meanng_roadSign[].

Learning of the knowledge of the category of road signs involved learning of the color coding of road signs. Color coding is connected with recognition of color of the sign. Recognition of color of the sign is a part of the meaningful decomposition of the sign into parts. During learning of a visual object from the signs category, the object is given as the object from the category of colour class. Color is coded according to the selected meaning of colors (e.g North American and Australian colors coding). A color schema is stored in the text file in the form of arrays: names[], color backgrounds[], and color symbols[]. During learning color is coded as a member of a color schema: white, black, red, yellow, orange, blue, green. Based on a shape of a background, a shape of a symbol, a background color and color of symbols of the visual object the object is assigned to one of the specific sign categories.

For learning of the category of road signs, the sample of the representatives (55 objects) of this category was selected. Examples of the objects used for learning from the category of road signs are shown in Fig. 8.32 and 8.33. The signs used for learning were transformed into the images 1024x1024 (16 colors). The member of the sign category is decomposed into parts, background and figure, based on both the visual and the meaningful decomposition. The road sign has also its colour code, that means the colour and shape indicate the category of the sign category. At first, the colours of the sings were coded as blue, green, yellow, orange, red, white and black. Next, the color coding was applied to both the figure and the background. As a result the signs were assigned to the colours category as follows: red-white (R_W), blue-white (B_W), black-white (C_W, red-blue (R_B), green-yellow (G-Y), red-blue-white (R_B_W) or red_yellow_black (R_Y_C). Then each sign was assigned to the background-shape category described by the symbolic name of the category. After learning, the objects from the road signs category are tested if they have the proper symbolic names. For example, the different backgrounds of the road signs were learned (see Fig 8.32 (a-c)) and the following symbolic names were obtained "K "-(Z1_B) Fig 8.32 (a), "M3_L6_Nsr_0k0k0k"-(Z2_B) Fig 8.32 (b), and "M4_L8_Msr_0k0k0k0k"-(Z3-B) Fig. 8.32 (c). Next, figures of the road signs Fig. 8.32 (i-o) were learned and the symbolic names were obtained in a similar way. For example, the symbolic name for the figure shown in Fig. 8.32 (i) is "Q2_u_u_L7_qqqaqaq_L3_R_L3_R"-"F1". Then the symbolic name of the learned figure is combined with symbolic names of the background and tested during understanding process. For example, the symbolic name of the objects (figure) shown in Fig. 8.32 (i) F1 was combined with backgrounds Z1_B, Z2_B and Z3_B and the following symbolic names were obtained:

„A1_Q2_u_u_K1_L7_qqqaqaq_L3_R_L3_R”-(Z1_B_F1),
„A1_Q2_u_u_M3_L6_Nsr_0k0k0k_L7_qqqaqaq_L3_R_L3_R”-(Z2_B_F1),
„A1_Q2_u_u_M4_L8_Msr_0k0k0k0k_L7_qqqaqaq_L3_R_L3_R”-(Z3_B_F1).

Similarly, for all combinations of backgrounds and figures the symbolic names were obtained. The obtained symbolic names were next marked as the road sign “RS” or not-road sign “NSR”. It should be noted that not all combinations of the road signs background and road signs figures “gives” the road sign. For example, combination of the road sign background (Fig 8.32 (c)) and road sign figure (Fig. 8.32 (i)) do not give the road sign (Fig 8.32 (f)). The proposed method of learning makes it possible to understand the difference among signs and facilitate understanding of their meanings. At the final stage of learning, knowledge of each learned member of the road signs category is stored as knowledge of the knowledge schema. The knowledge schema stores the colour coding and meaning of the sign. For example, the sign represented by symbolic name „A1_Q2_u_u_M3_L6_Nsr_0k0k0k_L7_qqqaqaq_L3_R_L3_R”-(Z2_B_F1), is coded as {color (R-Y-C), figure color (C)}, {meaning-warning sign W1-P}. The knowledge schema stores only basic meaning of the road sign. The full meaning of the sign and its interpretation is learned as the part of the knowledge of the Sign-Knowledge-Expert. During understanding process meaning of the sign is interpreted by the Road-Expert or the Driver-Expert and translated into the robots behavior during learning of the driving vehicle or moving on the street. In this book these issues are not discussed.

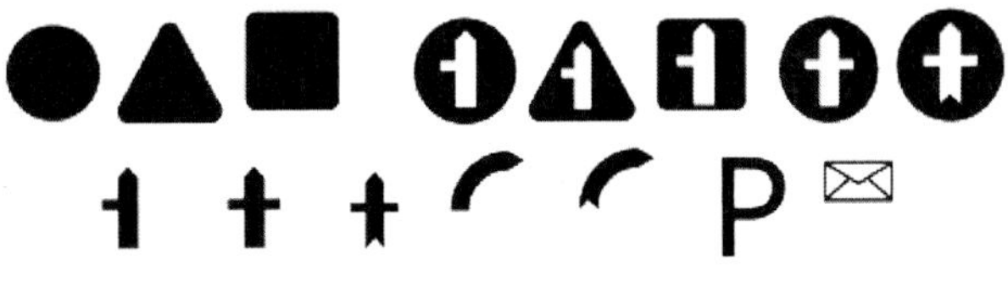

Fig. 8.32. Example of objects used for learning the category of the road signs

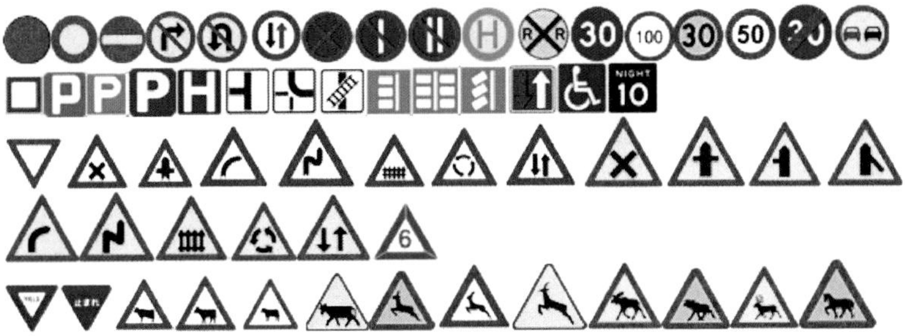

Fig. 8.33. Examples of objects used for learning of the road sign categories

8.1.2.5 Implementation of the New Skills – The Real World Objects Category

In this section learning of the knowledge of the selected categories of the real world objects category is presented. During learning of the knowledge of the category of real world objects, the following selected categories were learned: the category of the flat objects, the category of objects with flat parts and the category of the body revolution. In this book only learning of the knowledge of the knife category, which belongs to the category of objects with flat parts, in the context of learning of the parts decomposition, is presented. The knife category is the category derived from the category of objects with flat parts. The knife category is selected because objects from that category can be learnt from one characteristic view of the object. Learning of the knowledge of the knife category is to learn both visual and the non-visual knowledge of that category. The basic non-visual knowledge is learned during learning of the knowledge of the knowledge schema. The knowledge schema includes the name of the category and the definition of the category. Definition can be given in many different forms that show selected aspects of this category. For example, the definition of the type "used for" defines a knife as "knife is tool for {cutting, slicing, spreading, skinning, boning, paring, breaking, carving, peeling, sticking, trimming, mincing}". The definition of the type "consist of" defines a knife as a tool that consists of a blade and a handle. The definition of the type "made of" refers to a material category used for making knife "knife made of" {stainless steel, molybdenum stainless steel, high carbon steel, high carbon stainless steel}. In the case when parts of the knife are made from the different materials the definition is given in the form "handle made of {metal, wood, plastic}, blade made of {metal}". The definition of the type "type of" defines the specific category of knives such as "[bread slicer, bread knife, butcher knife, cheese knife, cimeter knife, cook's knife, cutlery knife, boning knife, paring knife, steak knife, breaking knife, carving knife, chef's knife, fillet knife, oriental knife, sandwich spreader, spreader, seafood knife, silex knife, pastry knife, peeling knife, scimitar knife, serrated knife, skinning knife, sticking knife, utility knife, trimming knife, cleavers, mincing knife]". From these specific categories the very specific categories are derived, for example, from the category of bread slicer the following categories are derived "[mundial bread, bread knife, Household Bread Knife, Superior Offset Bread Knife]". The super-specific categories derived from the very specific categories define the category of knives in terms of a size, a type of blade, a type of edge, a blade material and a handle material and color, for example, "[mundial bread slicer knife [10", curved blade, micro-serrated edge, stainless steel, black plastic]". The knife category has its main features such as *bladetype, size, edgetype, parttype* that can be used to derive very super-specific categories. Each main feature has its domain: for example, bladetype [straight, curved, curved semi stiff, narrow stiff, extra wide stiff], size[12", 10", 8", 71/2", 6"], edgetype [micro-serrated, wavy], parttype [onepart, twopart, manyparts], material [stainless steel, Rigid Molybdenum Stainless Steel, molybdenum stainless steel, high carbon steel, high carbon stainless steel].

Objects from the knife category used for learning of the visual knowledge of the knife category are members of the two perceptual categories: the silhouette category and the colour category. Objects from the colour category can be used to learn parts decomposition. However, in the experiment presented in this book the parts are learned independently from the silhouette category. This approach makes it possible to avoid errors during learning of small details. At first, the general visual concept of the knife is learnt based on the selected representatives. Next, the specific knife categories are selected and the visual concept of these categories is learnt in the context of the visual concept of its parts. It should be noted that learning for understanding is different from learning for recognition. In the understanding process not only knowledge used for recognition of the object is important one but also knowledge of the parts used in the assembling process and knowledge of the assembling process.

During learning, the objects from the shaded-objects category were transformed into the objects of the colour-objects category and next into the objects of the silhouette category. Fig. 8.34 shows examples of the objects from the shaded-objects category transformed into the objects of the silhouette category. Also, the objects from the line-drawing category were transformed into the objects of the silhouette category. Fig. 8.35 shows examples of the objects from the line-drawing category transformed into the objects of the silhouette category.

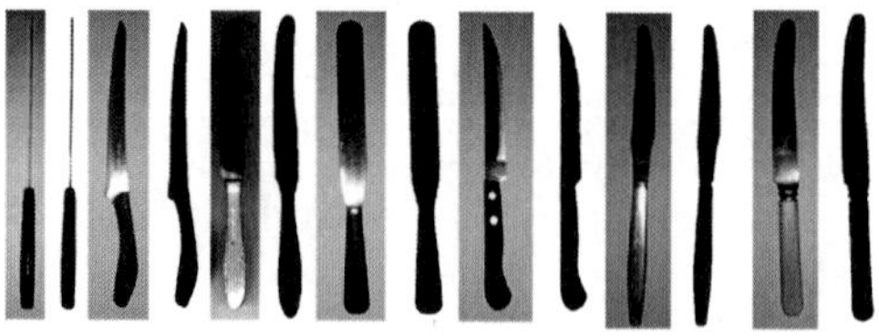

Fig. 8.34. Examples of the objects from shaded category transformed into the silhouette category

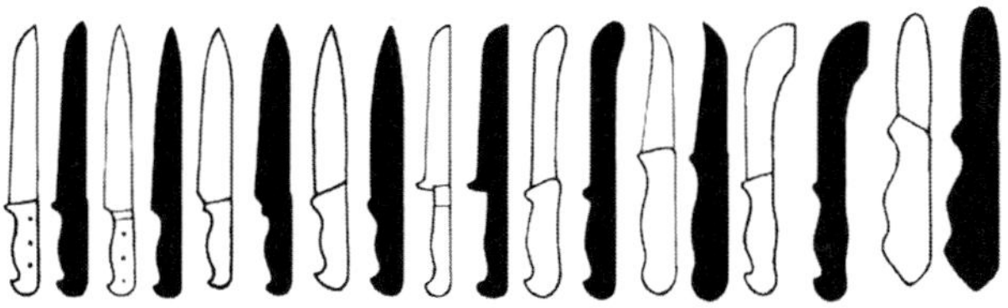

Fig. 8.35. Examples of the objects from drawing category transformed into the silhouette category

During learning of the part decomposition, parts obtained from the objects of the color-object category were transformed into the objects of the silhouette category and were learned as the independent visual objects. During learning, the object is divided into parts that are indicated by the different colors and after learning, the visual concept of these parts is stored by linking them with the visual concept of the object. The parts from which the object is assembled are learned independently as knowledge of the assembling process. Fig. 8.36. shows examples of parts used in the assembling process. In this book learning of the knowledge of the assembling process is not presented.

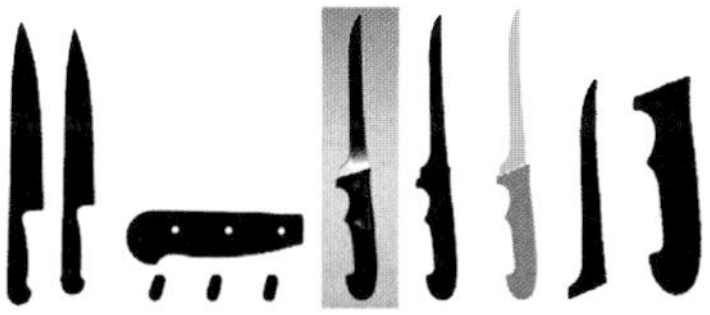

Fig. 8.36. Examples of learning of the parts

During learning of the visual concept of the knife category each object is, at first, transformed into the silhouette category. A prototype is learned by application of the method of learning by simplification of the complex object, in order to learn similar objects that are used during understanding process to understand of the similar objects. Learning by simplification of the complex object is applied when the object is of high complexity and there are no well elaborated image transformations for this type of object. The object is at first simplified and SUS starts to learn from the simple objects followed by the more and more complex one. Examples of the representatives used in learning process are shown in Figs. 8.37-8.40. The objects from the category of objects with flat parts are shown in Figs. 8.37 and 8.38. Fig. 8.40 shows the category of symmetrical objects that refers to the category of the body revolution objects. The interpretation of visual objects, shown in Fig. 8.40, will be given as the interpretation in terms of the category of objects with a flat part or as the interpretation in terms of the category of the body revolution object. For the purpose of learning of the visual concept of the knife category 170 objects from the knife category were selected. The complex object that is learned is decomposed into simple objects as shown in Fig. 8.39.

During learning process the new classes are established and the new image transformations are implemented.

Fig. 8.37. Examples of object used to learn knife category

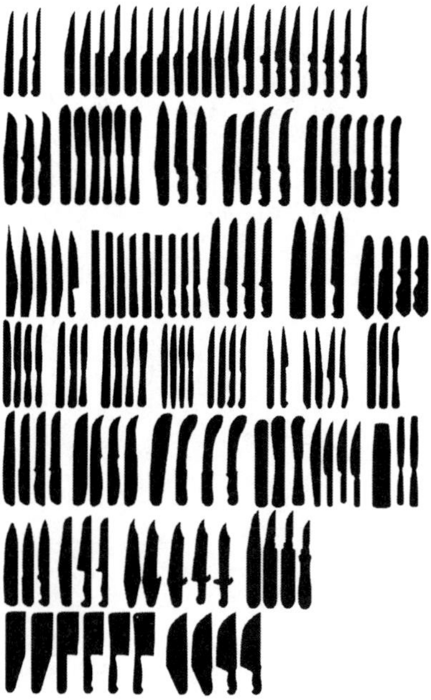

Fig. 8.38. Examples of objects used to learn a knife category

Fig. 8.39. Example of objects used in learning of parts decomposition

Knowledge of the category of symmetrical knives is learned in the context of the knowledge of the category of the body revolution objects. Examples of symmetrical objects used in learning are shown in Fig. 8.40.

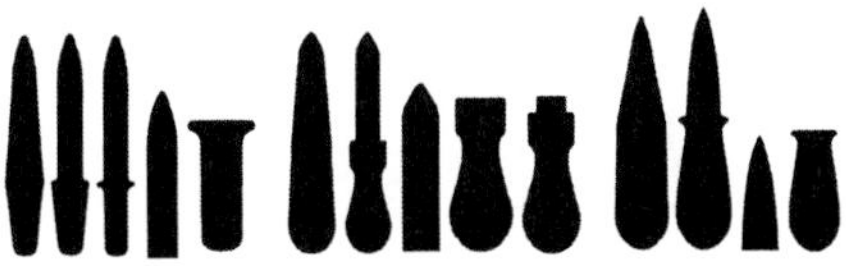

Fig. 8.40. Examples of objects that can be regarded as objects with flat part or objects of body revolution

The results of learning by application of the different categorical learning methods such as learning by the small alternation of the visual object (LSA), learning from the simple to complex (LSC), learning by simplification of the complex object (LSCO), learning from parts (LP), and learning parts decomposition (LPD) show that these methods give satisfactory to good results. These methods were applied to learn of the visual knowledge from the selected ontological categories such as the signs category, the letters category and the knife category. Application of these methods to learn of the visual knowledge of the selected ontological categories showed that each of the methods performed well. In the case when an object was well visible (without small sides, small angles and with regular curvilinear segments) learning of the visual knowledge gave very good results. The most difficult problem in learning is to cope with learning objects with small sides and curvilinear segments. This problem is regarded as a problem connected with visibility of an object and a perceptual ability of SUS to perceive an object. The perceptual ability of SUS to perceive an object is connected with the size of the image. The complex object can have some parts that are too small to be recognized by SUS.

One of the problems to be solved during learning of the visual objects is the difference in the symbolic names of visual objects perceived by SUS (perceived as the different objects) whereas they are perceived by a human observer as nearly identical. Fig. 8.41 illustrates this problem using as an example a cipher 6. The cipher 6 in Fig. 8.41 (a) is very similar to the cipher 6 in Fig. 8.41 (c). However, due to the small differences in the size of the "top" parts, the residuals are different and have the symbolic names that are quite different $Q^1[M^1[L^4]](M^1)$ for the object in Fig. 8.41 (b) and $Q^2[M^1[L^4]](Q^1[L^3](M^1))$ for the object in Fig. 8.41 (d). In the case, when this difference is "big enough" these two visual objects can be "seen" as different objects although the human observer very easily can "take them" as being the same object. In the case when this difference is "small" the "thin" part in the residue Fig. 8.41 (d) can be the source of the perceptual error. The very thin part (1-pixel) needs to be differently processed. This type of error is corrected during the reasoning process at the intermediate level of understanding. The correction process involves assumption that the visual objects are objects from a complex class. As we can see, both objects have the same symbolic names (symbolic representation) $C[Q^1[M^1](Q^1[L^3](M^1)),Q^1[M^1[L^5]](Q^1[M^1[L^4]](M^1))$ in notation of the complex class. This symbolic name can be used in recognition process in order to avoid errors in interpretation of the object. Knowledge of the correction process is stored as part of the knowledge of the Intermediate-Level-Name-Expert. This knowledge includes a symbolic name of a structural archetype, the symbolic names of both a concave and a complex class model and image transformations for processing object of the complex class.

Fig. 8.41. Cipher "6" as an example of difference in SUS visual interpretation

The perceptual ability of SUS to perceive the side of an object depends on the size of the object. In the case of an object from the convex polygon class, the object with the small side can be assigned to one of the classes: L^n, L^{n-1} or $\widehat{L}^n$. Similarly, in the case of an object from the concave polygon class, the object can be assigned to $Q^m[L^n]$, $Q^m[L^{n-1}]$ or $Q^m[\widehat{L}^n]$ class. For example, the object from the class $Q^2[L^6]$ shown in Fig. 8.42 can be assigned into one of the classes: $Q^2[L^6](2L_R^3)$, $Q^2[M^1[L^6]](2L_R^3)$, $\widehat{Q}^2[L^6](2L_R^3)$ or $Q^2[L^5](2L_R^3)$. Knowledge of the correction process or contextual knowledge can be used to select the proper visual interpretation. Knowledge of the correction process can be given in the form of the symbolic names of the complex classes. Similarly objects shown in Fig. 8.43 can be interpreted by applying knowledge of the correction process or the contextual knowledge.

Fig. 8.42. The different interpretation of object with a small side

Fig. 8.43. The different interpretation of the object with a small side

A polygonal object with the angle between sides close to 180 degrees can be interpreted as L^n or L^{n-1}. For example, the object shown in Fig. 8.44 can be interpreted as object from the following classes: L^6, L^5, or L^4, depending on the size of the object and the angle between sides of the object. In order to make interpretation more perceptually oriented, a new specific class is established $\text{Ł}^6(2)$ that shows that two angles are close to 180 degrees. The object assigned to this class can be interpreted as the object from the distorted L^4 class or as the object from the specific L^6 class. This interpretation is very important in the case when there is another evidence that the object can be from one of the classes: L^4 or L^6.

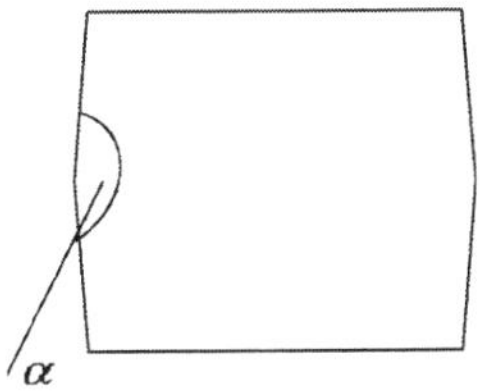

Fig. 8.44. The different interpretation of the object with a small angle (close to 0)

Perceptual ability to see curvilinear segments depends on the size of the curvilinear segment and the type of the curvilinear segment. The object with the curvilinear segment from the class M^1 can be assigned to one of the classes: M^1, L^n or $M^1[L^3]$ or $M^2[L^4]$. In order to reduce an error connected with perception of the curvilinear segment the new classes are established. For the small elongated object, the class M^1_T is established and for the elongated object, the class M^1_L is established. In order to find the proper symbolic name, knowledge of the shape classes is used during interpretation of the object at the intermediate level. Fig. 8.45 shows objects that can be interpreted as members of the curvilinear class. As we can see in the case when the object is small the unique interpretation is very difficult. For example, the object shown in Fig. 8.45 b can be interpreted as the object from one of the classes: $M^1[L^3]$, L^3 or M^1. Similarly, the object in Fig. 8.45 e can be interpreted as the object from one of the classes: $M^2[L^4]$, M^1 or L^4.

Fig. 8.45. The different interpretation of the object with curvilinear segment

Learning of the non-visual knowledge was restricted to learning of the knowledge of the categorical chain and knowledge of the knowledge schema. Learning of the non-visual knowledge will be a topic of a further study.

8.1.3 Implementation of the New Skills – The Category of the Sensory Objects

In previous Sections learning of the knowledge of the category of visual object was presented. In this Section learning of the knowledge of the category of the sensory object is presented. As it was described in Chapter 7, the sensory object is an object that is named based on the measurement of values of the different object attributes. Visual attributes such as color or shape does not supply enough information to obtain the reliable recognition and understanding of the sensory object. The category of sensory objects derived from the category of the visual objects, such as the minerals category or the rocks category, is called the category of visual sensory objects. A mineral is a naturally occurring solid chemical substance formed through geochemical processes, having characteristic chemical composition, highly ordered atomic structure, and specific physical properties. Naming of the object from the minerals category (recognition or classification) is to assign the name $n(v_M^i)$ one of the mineral categories v_M^i to the examined object o_i based on a set of measurements $m(a_i^k)$ and matching processes. The aim of the matching process is to assign an examined object to the mineral category v_M^i based on matching of the values of the attributes of the examined object $m(a^j)$ with the values of the attributes of the mineral category $val(a_i^j)$.

Naming of an object from the minerals category is part of the understanding of the objects from the minerals category. Understanding of the minerals categories is connected with understanding of the mineralogical texts and understanding of the geological categories. One of the aims of conduced research is to build the system of mineral understanding “Domeyko”. The system of mineral understanding “Domeyko” is further development of the system of minerals recognition implemented in Prolog [63], [136]. The system of minerals recognition was able to recognise the object (mineral) based on the values of the minerals attributes. The recently developed system of minerals understanding “Domeyko” performs understanding assuming that the examined object can be the object from any category of visual objects such as the rock category, the glass category, the plastic category, the ceramic category, or the wood category. In the case when there is the reliable information that the examined object is collected in the field, the system assumes that the object can be the object from the rock category or the mineral category.

The mineral category v_M^i is given by its name χ_i. The result of the naming process is assigning of the name of the category to the examined object. Proper naming of the examined object is vital because after naming, all knowledge concerning minerals category v_M^i given by the name χ_i is accessible through the Mineral-Understanding-Expert. The first stage of minerals' understanding is based on the measurement of physical attributes of minerals, such as the size $a(L)_j^{Ph}$, the specific gravity $a(G)_j^{Ph}$, the hardness $a(H)_j^{Ph}$, the color $a(C)_j^{Ph}$, the streak $a(S)_j^{Ph}$, the cleavage $a(K)_j^{Ph}$, the diaphaneity $a(D)_j^{Ph}$, the luster $a(L)_j^{Ph}$, the habit $a(H)_j^{Ph}$, the tenacity $a(T)_j^{Ph}$, the fluorescence $a(F)_j^{Ph}$, the magnetism $a(M)_j^{Ph}$, the radioactivity $a(R)_j^{Ph}$, or the piezoelectricity $a(P)_j^{Ph}$. The quantitative physical attributes have their values given as the numerical values, discrete or continuous. The discrete attribute is the quantitative attribute whose value is countable. The hardness $a(H)_j^{Ph}$ is a discrete attribute that has value $val(a(H)_j^{Ph}) \equiv \{1,...,10\}$. The magnetism $a(M)_j^{Ph}$ is a discrete attribute that has value $val(a(M)_j^{Ph}) \equiv \{0,1\}$. The specific gravity $a(G)_j^{Ph}$ is a continuous attribute that has value in the interval 1-26. The colour $a(C)_j^{Ph}$, the streak $a(S)_j^{Ph}$, the cleavage $a(K)_j^{Ph}$, the diaphaneity $a(D)_j^{Ph}$, the lustre $a(L)_j^{Ph}$, the habit $a(H)_j^{Ph}$ are categorical attributes whose value are categories c_k^j. The colour attribute $a(C)_j^{Ph}$ has its value $val(a(C)_j^{Ph}) \equiv \{c_1^C,...,c_N^C\}$.

Measurements $m(a_i)$ of the attributes a_i such as physical, chemical, optical and Y-Ray attributes can be obtained by applying the different measurement apparatuses with the specification of the measurement process $A(a_i)$ $\{A^S, A^K, A^M .A^R\}$, where A^S is a way of preparation of the special sample of minerals, A^K is knowledge of the method of measurement, A^M is the prescription of steps of the measurement method and A^R are measurement results. The values of the quantitative attributes differ depending on the type of the attribute measured. For example, the hardness $a(H)_j^{Ph}$ has values in the interval 1-10, $val(a(H)_j^{Ph}) \equiv \{1,...,10\}$, the magnetism $a(M)_j^{Ph}$ has values in the interval 0-1 $val(a(M)_j^{Ph}) \equiv \{0,1\}$ or the specific gravity $a(G)_j^{Ph}$ has values in interval 1-20.

Measures of the categorical attributes such as the colour $a(C)_j^{Ph}$, the streak $a(S)_j^{Ph}$, the cleavage $a(K)_j^{Ph}$, the diaphaneity $a(D)_j^{Ph}$, the luster $a(L)_j^{Ph}$ or the habit $a(H)_j^{Ph}$, are obtained and used at the first stage of minerals' recognition. The colour $a(C)_j^{Ph}$ and the streak $a(S)_j^{Ph}$, similarly as other categorical attributes, have their values given by the name of the colour categories $val(a(C)_j^{Ph}) \equiv \{c_1^C,...,c_N^C\}$. The values of the colour attribute $val(a_i^j)$ that are part of mineralogical knowledge are measured on the subjective scale and by this it can not be very precise. Also subjective evaluation of the colour categories $\{k_i^C\}$ as a measurement $m(a_i^k)$ of colour predicates can be inaccurate. During recognition these inaccuracies can cause serious error during naming process and can lead to assigning the wrong name to the examined object. In order to avoid this sort of errors, each colour category $\{k_i^C\}$ and $\{c_j^C\}$ is transformed into the basic colour scale and then into a fuzzy scale of colours.

Typically the colour categories $\{c_i^C\}$ are established based on the subjective criteria [137]. As the result, matching colours during process of recognition and understanding of the objects can be very difficult. To facilitate matching, the colour categories $\{c_i^C\}$ and $\{k_i^C\}$ are transformed into the basic colour categories $\{c_i^{CB}\}$ and $\{k_i^{CB}\}$. For each mineral, values of the colour attributes $val(a(C)_j^{Ph}) \equiv \{c_1^C,...,c_N^C\}$, in the form of the colour categories $\{c_i^C\}$, are established during mineralogical scientific research. Each mineral is described by value of one or more then one colour categories $\{c_1^C,...,c_N^C\}$. These categories are printed in scientific papers or books as part of the mineralogical knowledge. However, there is a lack of agreement on proper names of colour categories and many scientific books give different information about the values of the colour attributes by assigning the different colour categories to the same mineral. Because the colour categories $\{c_i^C\}$ are established based on the subjective criterion they can cause an error during matching process and also subjective evaluation of the colour categories $\{k_i^C\}$ can be inaccurate and by this can be source of errors. For example, the category "deep green" or "emerald" can be used to denote "green" object and used to describe mineral such as malachite, whereas an examined object can be perceived as "blue green".

A colour category (colour name) can be expressed by one term, two terms, or three terms. A colour name can be given by the name of basic colours: colourless, white, black, grey, brown, yellow, red, blue and green, or by the name of shadows categories e.g. greenish, grayish. A colour name can be given in the form of names of two colours e. g. blue green, or in the form of name of shadow category and the name of basic colour e.g. greenish blue or in the form of names of three colours e.g. violet steel gray. A colour name can be formed by adding colour marker such as bright, pale, light, nearly, dark, deep, shining e.g. bright blue. A colour name can refer to the category of objects that has characteristic colours; the category of fruits: lime, lemon, citron, cherry, raspberry, melon, peach, plum, the category of food: olive, tomato, wheat, grass, honey, milk, the category of flowers: rose, cornflower, dodger, lilac, orchid, the category of things: firebrick, brick, slate, the category of environment: sea, sky, spring, snow, dirt, the category of animals: peacock, salmon, liver, coral, flesh, the category of minerals: aquamarine, emerald, ruby, turquoise, the category of materials: lead, tin, bronze, brass, gold, steel, copper or the category of colour itself: vermilion, ultramarine, alizarin crimson, magenta, indigo, or beige. The colour names are used to learn the coding categories needed during understanding of the mineralogical texts.

In order to avoid misinterpretation the colour categories $\{c_i^C\}$ and $\{k_i^c\}$ are transformed into the basic colour categories $\{c_i^{CB}\}$ and $\{k_i^{CB}\}$. The basic scale consists of the basic colours: colourless, white, black, grey, brown, yellow, red, blue, green. For example, by applying this scale to colours such as "pale green" or "deep green" these colours are transformed into the colour "green". Similarly, the name consisting of two words such as blue-green is transformed into two colours "blue" and "green" and the name such as 'emerald' is transformed into the "green" colour.

colour	Basic colour	Basic colour
brown red	Brown	red
emerald	Green	green

The fuzzy scale is designed taking into account the similarities among colours. For example, the colour blue can be confused with the colour green or violet but not with red or yellow. The fuzzy notation shows the colour and its membership function. For the colour blue the membership function has following expression: BLUE B(1), G(0.9) V(0.9) C(0) W(0) Bl(0) Gr(0) Br(0) Y(0) R(0).

The fuzzy scale is implemented as an array of numbers:

colour	B	G	V	C	W	BL	Gr	G	Br	Y	R
blue	1	0.9	0.9	0	0	0	0	0	0	0	0

During naming of an examined object the colour categories $\{c_i^c\}$ and $\{k_i^c\}$ are transformed into the fuzzy scale and as the result of naming process a set of possible names $\{ n_0(v_M^i)[p_0],...,n_j(v_M^i)[p_j],...,n_K(v_M^i)[p_K]\}$, where $[p_j]$ are possibility coefficients, are obtained. Application of the fuzzy scale make it possible not to exclude the name of mineral for which value of the mineral attribute is erroneously measured. For example, let assume that an examined object is a member of the cuprite category that is described by following colours: brown red, purple red, red, and black. When colour of an examined object is evaluated as the deep red, the examined object will be excluded from being named as a member of the cuprite category. In the case when the basic scale of colours is used, the cuprite has following colours: brown, red, purple, black and the colour of examined object is red, so the examined object will be named as member of the cuprite category. In the case when the colour of examined object will be evaluated as a grey only an application of the fuzzy scale does not exclude an examined object for consideration as a member of the cuprite category. Similarly, other categorical attributes such as the habit or streak are transformed into the basic scale and next into the fuzzy scale. The habit category is composed of the specific habit categories such as crystal face $a(Hf)_j^{Ph}$, the crystal form $a(Hc)_j^{Ph}$ and the crystal aggregate $a(Ha)_j^{Ph}$ and all specific categories are transformed into the basic scale and next into the fuzzy scale. The measurement of numerical attributes such as hardness can be easy to obtain, whereas measurement of the specific gravity requires the use of a special apparatus, Jolly balance.

During naming of an examined object the categorical categories $\{c_i^x\}$ and $\{k_i^y\}$ are transformed into the fuzzy scale and as the result of naming process a set of possible names $\{n_0(v_M^i)[p_0],...,n_j(v_M^i)[p_j],...,n_K(v_M^i)[p_K]\}$, where $[p_j]$ are possibility coefficients, are obtained. The naming is based on reasoning process described in [1]. The algorithm of the first stage of naming process is given as follows:

For i=1 to N Input k_i^y
Transform k_i^y *into fuzzy scale* $k_i^{yF} = F(k_i^y)$
For j=1 to M
Transform c_j^x *into fuzzy scale* $c_j^{xF} = F(c_j^x)$
For j=1 to M
For i=1 to N
Match $n_k(v_M^j)[p_k] = k_i^{yF} \otimes c_j^{xF}$

where $n_k(v_M^j)[p_k]$ denotes result of matching process $k_i^{yF} \otimes c_j^{xF}$, and $p_k > \varepsilon$, and ε is the assumed threshold, N, M are numbers of the k_i^y and c_j^x categories.

In the second stage of naming process the chemical attribute a_j^{Ch} such as reactivity to dilute acids $a(D^A)_j^{Ch}$, reactivity to dilute water $a(D^W)_j^{Ch}$ or the flame test attribute $a(F)_j^{Ch}$ are used. These attributes are measured by performing appropriate tests. For example, the wet tests consists in checking to see whether a mineral is water soluble, then whether it is soluble in hydrochloric acid, diluted and cold or hot, or concentrated and cold or hot, and finally in nitric acid and in aqua regia. Flame tests consist in heating a mineral in a gas flame (about 1000°C; 1832°F). These tests do not require the complex measurement tools. The algorithm of the second stage of naming process is given as follows:

For k=0 to K

$N_h = n_k(v_M^j)[p_k] \in \{a(X)_j^{Ch}\}$, where N_h is group of minerals that have the value of attribute $a(X)_j^{Ch}$

For each group N_h perform test $N_h^T = T[a(X)_j^{Ch}]$
For h=1 to H
If $N_h^T = 1$ $U = U \bigcup \{n_h^T\}$, where $\bigcup$ denotes the sum of sets $\{n_h^T\}$
If $|U| = 1$ than $n = n_h^T$ else perform further tests.

If there are more than one name in a set $\{n_h^T\}$, $|U| > 1$ the third stage of naming process is started.

In the third stage of naming process the minerals in a set U are differentiated based on any available information that can help in selection of the most possible candidate that fulfil criterion if $|U|=1$ than $n=n_h^T$. Information such as the locality, where an examined object was collected or name of associated minerals can be used to select among possible names.

In the fourth stage of naming process the measurement of instrumental method spectroscope attribute $a(I^S)_j^{Ch}$, instrumental method infrared spectrograph attribute $a(I^{IR})_j^{Ch}$, or the X Ray diffraction analysis attributes a_i^X, such as X-Ray powder diffraction $a(P)_i^X$ are recommended. Also, the optical attributes a_i^P: the type $a(T)_i^P$, RI coefficient $a(R)_i^P$, 2V $a(V)_i^P$, maximum birefringence $a(B)_i^P$, surface relief $a(S)_i^P$, dispersion $a(D)_i^P$, Gladstone-Dale index $a(G)_i^P$, biaxial $a(b)_i^P$, pleochroism $a(P)_i^P$ can be used.

Measurement of other chemical or optical attributes requires applying the specific apparatuses. Instrumental analysis, which is the best methods used for small samples, uses electron microprobe that measure extremely small quantities of inorganic material. Spectroscope is used for qualitative analysis of the elements contained in a mineral on the basis of the presence of absorption lines in the light which has passed through the specimen and then been dispersed by a prism. Infrared (IR) spectrograph measures vibrational frequencies of atomic groups (especially of water or the hydroxyls) present in a mineral. Differential thermal analysis (DTA) measures loss of water and other gases. X-ray powder diffractometer is used to identify minerals and to ascertain their crystal structure. In order to perform complex analysis knowledge of the measurement by application of the special methods and different measurement tools is required. This knowledge is part of the knowledge of the interpretational script of the dictionary-text described in Section 7.3.3. The knowledge of different measurement tools represented as $A(a_i)\ \{A^S, A^K, A^M.A^R\}$, where A^S is a special sample of minerals, A^K is knowledge of methods of measurement, A^M represents consecutive steps of the selected method and A^R the measurement results, is stored as part of the Mineral-Dictionary-Script-Expert. The measurement results A^R are stored in the text format. For each established mineral category the values of attributes of the mineral category $u(a_i^j)$ are measured and presented in the different forms (mineralogical data) in scientific books. Mineralogical data are also available on the www presented in the different formats. Examples of two different forms of data for minerals cookeite and adamite' are presented below. The Mineral-Data-Expert transforms the data into SUS format and stores it in the text format. During naming process mineralogical data are used by The Mineral-Naming-Expert to match the measurement results A^R that are obtained during measuring attribute a_i of an examined object by applying the measurement tool $A(a_i)$.

Cookeite Crystallography
Axial Ratios: a:b:c =0.5744:1:3.2138
Cell Dimensions: a = 5.13, b = 8.93, c = 28.7, Z = 4; beta = 98.75° V = 1,299.47
Crystal System: Monoclinic - PrismaticH-M Symbol (2/m) Space Group: P 21/a
X Ray Diffraction: By Intensity(I/Io): 2.315(1), 3.52(0.9), 4.7(0.9),
Optical Properties of Cookeite
Gladstone-Dale: CI meas= 0.044 (Good)
KPDcalc= 0.2185, KPDmeas= 0.2185,KC= 0.2286
Optical Data: Biaxial (+), a=1.572-1.576, b=1.579-1.584, g=1.589-1.6, bire=0.0170-0.0240
Crystallography of Adamite
Crystal System: Orthorhombic
Class (H-M): mmm (2/m 2/m 2/m) - Dipyramidal
Space Group: Pnnm {P21/n 21/n 2/m}
Cell Parameters: a = 8.304Å, b = 8.524Å, c = 6.036Å
Ratio: a:b:c = 0.974 : 1 : 0.708
Unit Cell Volume: V 427.25 Å³
Z: 4
Optical Data of Adamite
Type: Biaxial (+/-)
RI values: nα = 1.708 - 1.722 nβ = 1.742 - 1.744 nγ = 1.763 - 1.773
2V: Measured: 78° to 90°, Calculated: 74° to 84°
Maximum Birefringence: δ = 0.055
Surface Relief: High
Dispersion: strong r > v or r < v
Pleochroism: Weak

The complex mineralogical data such as data of X-Ray Powder Diffraction can be given in graphical form or as the measurements of the characteristic points. Examples of the X-Ray Powder Diffraction data are shown in Fig. 8.46.

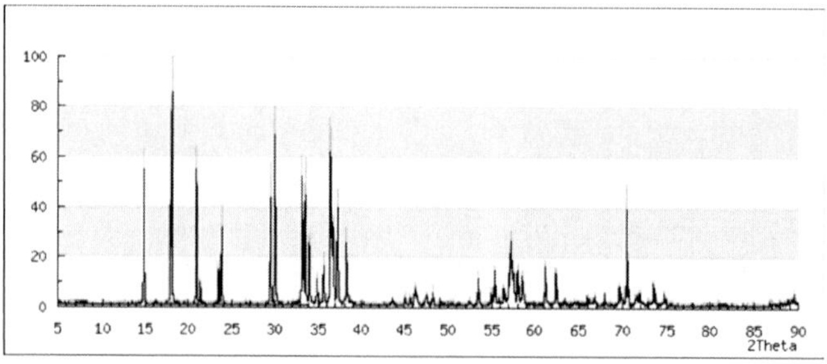

Fig. 8.46. X-Ray Powder Diffraction Data

The X-Ray Powder Diffraction data are stored in the text file in the following form:

X-Ray Powder Diffraction: d-spacing, Intensity

5.944, (60), 4.897, (90), 4.242, (60), 2.698, (80), 2.448, (100), 1.608, (80)

The algorithm for naming mineral object is implemented as a part of the Mineral-Nnaming-Expert.

8.1.4 Implementation of the New Skills – The Procedural-Form

In this Section learning procedural-form is described. Learning of the procedural-form is connected with learning of the problem solving skills that are used to solve educational tasks. Learning of the new problem solving skills is concerned with learning of the new methods of problem solving and implementing of the new methods of processing and storing of acquired knowledge. Learning of the problem solving skills involves learning of the knowledge and mechanisms how to apply acquired knowledge to obtain the solution of a problem (text-task).

Learning of the basic-form and the procedural-form is connected with learning of the problem solving skills. Some problems can be solved by applying several solution methods and learning of the different solution methods for the same problem can be useful for finding solution to similar problems. Learning starts with the more basic problems and progress to those that are more difficult problems. At first, the text-tasks are grouped into groups that have the same basic meaning. Each text-task T_i^T is transformed into the basic-form $B_j(T_i^T) = \beta_j$. Next, the text-tasks are grouped into groups that have the same basic-form $T_i^T \in T_i^T(\beta_j) if B_j(T_i^T) = \beta_j$. These groups are arranged according to the difficulty level. Each group of tasks $T_i^T(\beta_j)$ is further grouped into groups that have the same query-form $T_i^T \in T_i^T(\beta_j(\theta_k)) if \Theta_k(T_i^T(B_j)) = \theta_k$. For each basic-form $B_j(T_i^T) = \beta_j$ the procedural-form $P_j(T_i^T)$ is learned. For each group $T_i^T(\beta_j(\theta_k))$ the query-form $\Theta(T_i^T)$ and the explanatory script $J_j(T_i^T)$ are learned. Each basic-form can have one or more than one procedural-form depending on the number of solution methods used to solve the task represented by the basic-form. The procedural-form is used in the final stage of understanding of the text-queries or the text-task; this form is essential in finding of the solution. Learning of the procedural-forms is concerned with learning of the different methods of finding of the solution of the specific tasks.

Methods employed in finding of the solution of the task can be divided into two classes: computational methods and searching methods. The computational method is based on translating the text (linguistic form) into a mathematical model (expression). The simplest computational methods employ the simple mathematical operations. For example, the text-task "What is difference between 36323 and 35323", is translated into the basic-form "compute difference a1 and a2" and the procedural-form “Data a1, a2 formula F1: r=a2-a1. The similar task “how much greater than 5291 is 5691", has the basic-form “how much greater than a1 is a2” and the same procedural-form “Data a1, a2 formula F1: r=a2-a1”. Similarly, the statistical task “what is mean of these numbers 10 2 4 5 6 0 9 9 " is translated into the basic-form “Compute mean” and into the procedural-form “Data [a(1), ,a(n)] compute mean F1: m=(a(1)+...+a(n))/n”.

The procedural-form of physical text-task reflects the uniqueness of physical domain. In domain such as physics all measurements are expressed in units and there is a need to transform the measurement units from one to another. For example, to transform temperature given in Kelvin degrees into Celsius degrees, the formulae such as $T_{Kel} = \frac{5}{9}(T_{Far} + 459.67)$ is used. These computations are carried out by invoking Solve-Unit-Expert that implements unit transformation.

For more complex physical-text-task, the procedural-form describes steps of computation in the form of the concise description or algorithm. For example, for the text-task translated into the basic-form “object O move with speed v [m/s] in time t [s]. Find s” the procedural-form is given as “Data: v [vU], t [tU] Find: s, Convert v[vU]=v[m/s], t[tU]=t[s]”. Defining equation is: v=s/t. Solving equation for s: s=v*t. Inserting data s[m]=v[m/s]*t[s]”.

Some physical-text-tasks require transformation of the formula in order to obtain the proper form for computing required variable. For example, finding a formulae to compute force F, the kinetic energy U or the magnetic field B, is straightforward by applying an appropriate formulae $F = ma$, $U = \frac{1}{2}mv^2$, or $B = \frac{\mu_0 I}{2\pi r}$. In the case when a variable I is to be computed by applying formula $B = \frac{\mu_0 I}{2\pi r}$, the formula $B = \frac{\mu_0 I}{2\pi r}$ need to be transformed as $B = \frac{\mu_0 I}{2\pi r} \Rightarrow I = \frac{2\pi r B}{\mu_0}$. The Procedural-Form-Expert implements transformation in the following steps: At first $B = \frac{\mu_0 I}{2\pi r} \Rightarrow 1 = \frac{\mu_0 I}{2\pi B}$ next $1 = \frac{\mu_0 I}{2\pi B} \Rightarrow \frac{1}{I} = \frac{\mu_0}{2\pi B} \Rightarrow I = \frac{2\pi B}{\mu_0}$ or invokes the Math-Procedural-Expert to perform symbolic computation by applying *Mathematica* software.

The physical-text-task that requires solving complex mathematical problems, such as solving differential equations, is implemented as the Mathematical-Procedural-Expert. The Mathematical-Procedural-Expert implements the different methods that make it possible to solve both the numerical and symbolic mathematical problems. The Mathematical-Procedural-Expert invokes Math-Procedural-Expert to perform computation by applying *Mathematica* software. The cooperation of the Math-Procedural-Expert with *Mathematica* via MathLink creates the possibilities to utilize capabilities of *Mathematica* to perform the symbolic computations. MathLink is an open interprocess communication protocol that allows an external program to call *Mathematica* or to be called by *Mathematica*. MathLink is a high-level communication standard by which an external program can communicate with *Mathematica* kernel. Fig. 8.47 shows the class CMatlinkExpert that implements methods InvokeMathLink(), OnInitialUpdate(), CheckError (), OpenWMFFile() that made connection between *Mathematica* kernel and Generating Expert. The method Evaluate() reads the protocol sent by Mathematical Procedural Expert, translates it to the *Mathematica* expression and sends it to the *Mathematica* kernel. The method SendToDisplay() is sending back the results of the *Mathematica* calculation or graphical data to the Math Procedural Expert. The method DoDisconnect() disconnects the Generating Expert and destroys the CMatlinkExpert.

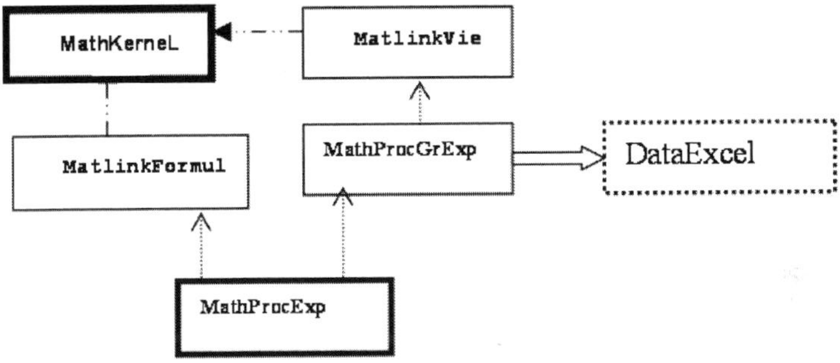

Fig. 8.47. Schema of communication between SUS and Mathematics

The Math-Procedural-Expert is sending message to the MathKernel in the form of a query. A query is given in terms of the *Mathematica* expression e.g. PolarPlot[2A*Cos[t]+2A*Cos[3t-B],{t,0.00001,2*Pi},AspectRatio->Automatic, Axes->False]. The query can be given in the form of the name of the geometrical curve e.g. "draw oblique trifolium" and the result can be presented in the form of the visual objects shown in Fig. 8.48. Depending on values of parameters A and B, different shapes of the curve are obtained.

Fig. 8.48. A curve of oblique trifolium obtained for the parameters 1. A=1, B=0 2. A=1 B=0.5

For the text-task formulated as: "find solution for the differential equation dy/dx+y=1", the Math-Procedural-Expert translates it into Mathematica expression DSolve[y'[x]+y[x]==1, y[x],x] and sends it via MathLink to MathKernel. The result of solving this equation is send back to Math-Procedural-Expert in the form {{y[x]->1+e-xC[1]}}. The Math-Procedural-Expert translates the solution into required form. In addition to the solution the Math-Explanatory-Expert can generate explanation concerning constant C[1] based on the explanatory script.

The Physical-Procedural-Expert cooperates with the Mathematical-Procedural-Expert during solving the mathematical problems. The simple mathematical formula such as "s=s1+s2=v1*t1+v2*t2" is solved by the Physical-Procedural-Expert by invoking the method SolveMathProblem(). Application of Mathematica implemented as the part of the Math-Procedural-Expert and invoked by MathLink supply the very good mathematical skills to SUS. These mathematical skills can be used to solve very complex mathematical text-task.

A text-task can be solved by applying an algorithm instead of a simple formula. A procedural-form describes an algorithm that is implemented as a part of the Algorithmic-Procedural-Expert. For example, the text-task "which number has smallest value 0.069 0.2 0.08 0.101" is translated into the basic-form "Find smallest number" and into the procedural-form in the form of the simple algorithm "Data [a(1), ,a(n)] find a: for all a(i), a(i)<=a".

The text-task used in intelligence tests are usually easy to understand but the solution cannot be found by applying the simple rule or formulae. The text-task such as "what is next number in this pattern 29 26 23 20 17" or "which number is missing from this number sequence 20 23 26 ? 32" are examples of the text-task that are solved using heuristics. The text-task "what is next number in this pattern 29 26 23 20 17" is translated into the basic-form "Find next number a1 a2 a3 a4 a5 ?" and next is translated into the heuristic procedural-form: "Data [a(1), ,a(n)] find a(n+1) by using methods H1: r(i)=a(i+1)-a(i), if r(1)= r(2)=...=r(n-1) than a(n+1)=a(n)+r H2: d(i)=a(i+1)/a(i) if d(1)= d(2)=...=d(n-1) than a(n+1)=a(n)+d, ..., Hn. The heuristic procedural-forms are implemented as the part of the Heuristic-Procedural-Expert. If needed, new heuristic methods can be added.

Problem solving skills that SUS acquired during learning of the different test-tasks depends on the learning process. Learning problem solving skills starts with the more basic problems and progress to those that are more difficult. The Procedural-Experts implements the different problem solving methods and the different methods of solving computational tasks. The Algorithmic-Procedural-Expert that implements algorithms needed to solve complex problems from computational geometry or graphs theory make it possible to apply these skills in solving text-task in many different scientific domain. The Heuristic-Procedural-Expert that implements the different heuristic methods supplies the very good computational skills to solve text-task that need to apply the non typical method to find the solution.

Finding of the solution of the text-task by employing searching methods is based on transformation of the text-task into a basic-form and the procedural-form. The basic-form is expresses in the form of the searching query. A procedural-form that is associated with a searching query describes steps needed to apply a searching method to find required facts. During learning process the Search-Procedural-Form-Expert implements searching algorithms and stores data in the form easily accessible by these algorithms. For example, learning of the procedural-form $P(o\_in\_X)$ is to select an object o_k, and next to learn knowledge at each categorical level represented by a container X^i_j. By selecting the categorical level (container) X^1_j all objects in this container are learned. By selecting the next categorical level (container) X_j^2 all objects in this container are learned. As the result, learned facts are stored in the form that is easily accessible by simple searching methods. The simple learning method, which requires vast amount of storage space guarantees a very simple mechanism of checking the correctness of learned facts. Learned knowledge is stored in an excel format during learning process and is converted into a text format when one of the search methods is applied. The excel format is used for an easy access to data during learning process. Data are accessed from Excel through Automation that is part of COM technology. The DataExpert loads Excel, creates a workbook and reads from and writes to cells from the active worksheet. Controlling Excel is like controlling an MFC Automation Component [124]. Proposed learning methods supply the easy way of checking of the correctness of learned facts and fast and easy access during understanding process. The excel format is used mostly during learning process, whereas the text format is used during understanding process. A procedural-form that is associated with a searching query is used in geographical tasks and is described in more details in Section 8.2.4.7.

The simplest problem formulated as a physical-text-task, is to find known quantity e.g. the Coulomb constant or gravitational constant. The procedural-form specifies stored knowledge format and describes the searching methods.

physical_constant	Symbol	number	Units
speed_of_light_in_free_space	c	2.997924*10^8	m/s

The problem of finding of the physical constants is carried out by invoking the Search-Physical-Constant-Expert.

8.2 Knowledge Implementation – Implementation of the New Knowledge

As it was described, learning of the new skills and new knowledge are processes that are strictly connected. Learning of the new knowledge is connected with learning of the knowledge that is used during understanding of the text. Learning of the knowledge and skills connected with understanding of the text will be called the learning of the new knowledge, whereas learning of the new knowledge and skills connected with understanding of the visual or sensory object will be called the learning of the new skills. Learning of the new skills was presented in Section 8.1. Learning of the knowledge of the visual object can be seen as learning of the specific perceptual skills to acquire data from an image and because of this learning of the knowledge of the visual object was presented in Section 8.1 as learning of the new skills. Knowledge learned by SUS is divided into two categories: the category of visual knowledge, and the category of non-visual knowledge. Knowledge associated with visual objects and sensory objects includes, the categorical chain, knowledge network of the categorical chains, knowledge schemes, knowledge categories, the model and theories. Knowledge associated with the text category includes: the categorical chain, the knowledge network of the categorical chains, the coding categories, the query-form, the basic-form, the procedural-form, the explanatory script, the interpretational script, and the categories of knowledge. Learning of the procedural-form was presented in Section 8.1.4. In this Section learning of the new knowledge is presented. The tests used for examining SUS performance were selected from the textbooks [138], [139], [140], crossword puzzles and AIM (Achievement Improvement Monitor) tests [141].

8.2.1 Implementation of the New Knowledge – Categories

In this Section learning of categories is described. Categories are basic components of acquired knowledge by SUS. As it was described in the previous chapters, the knowledge categories κ_i and the ontological categories of the visual object v_i are established based on existing knowledge. The coding categories c_i are related to both the knowledge categories κ_i and ontological categories of the visual object v_i, and are used during learning of the query-form $\Theta(T)$. The coding categories are learned during the iterative learning process and the association of the category c_i^j with a given word w_i, denoted as $\Re(w_i) \triangleright c_i^1,...,c_i^H$, can be changed during learning process. Knowledge of the basic interpretational level of SUS is stored in the form of categorical chains of the ontological categories of the visual object.

8.2.1.1 The Category of the Visual Objects

The categories of the visual objects are represented by the categorical chain. The categorical chain described in Chapter 6 is used during the first stage of understanding of a visual object. The perceived object, after naming, is interpreted as one of the category of the visual object. The categorical chains are linked and form a very complex structure of the knowledge that is used to interpret the visual object. The Categorical-Chain-Expert implements the categorical chains of the visual objects as the data "Catchainexicon.txt" stored in a text format and searching procedures for accessing the data. For example, the categorical chain:

$$O \triangleright \langle v_{Sig} \rangle \triangleright \langle v_{2DSig} \rangle \triangleright \langle v_{SymS} \rangle \triangleright \langle v_{CroS} \rangle \triangleright \langle v_{Lat}, v_X, v_{Pat}, v_{Pap}, v_{Lor}, v_{Mal}, v_{Cel}, v_{ChR}, . \rangle$$

is implemented as a set of categories and stored as follows: "Latin cross, CatLatCross, CatCross, CatSymSign, Cat2DSig, CatSign;" and the mineral-category given by the categorical chain:

$\langle \Pi \rangle \triangleright \langle \sigma_{El} \rangle \triangleright \langle v_{ReO} \rangle \triangleright \langle v_{Ear} \rangle \triangleright \langle v_{NLiv} \rangle \triangleright \langle v_{NMan} \rangle \triangleright \langle v_{Min}, v_{Rock}, . \rangle$ is implemented and stored as follows: "malachite, CatMin, CatNoMan, CatNoLiv, CatErthSize, CatREAlObj;".

During understanding process, the categorical chain is represented as an array of categories. During naming process, the Naming-Expert invokes the Categorical-Chain-Expert to supply the basic knowledge concerning the visual categories.

The Categorical-Chain-Expert implements the method that reads the text file with learned categorical chain and utilizes the categorical chain to reason about an examined object. The linked categorical chains are used during a categorical reasoning. The categorical reasoning is based on 'moving' through the linked chain categories. During the categorical reasoning, the Categorical-Chain-Expert invokes the Categorical-Chain-Reasoning-Expert to reason about the visual categories connected with the category of the perceived object.

An important category that is derived from the ontological category of the visual object is the part category. Learning of the categories such as the part category is conected with learning of the specific categories derived from the categories of visual objects. The part category is an auxiliary category that can be derived from any part of the categorical hierarchy. For example, the trees-category consists of the different specific part categories given by the following chain: $..\langle \nu_{Pla} \rangle \triangleright \langle \nu_{Tre} \rangle \succ [\tau_{Rot}, \tau_{Stm}, \tau_{Lef}, \tau_{Flw}, \tau_{Frt}, \tau_{Sed}]$. Each part category, such as the roots-category τ_{Rot}, the stems-category τ_{Stm}, the leaves-category τ_{Lef}, the flowers-category τ_{Flw}, the fruits-category τ_{Frt}, and the seeds category τ_{Sed} refers to the parts of a tree. The Part-Expert implements the categorical chain of the part category in the form of the data stored in a text format in file Partchainexicon.txt. For example, the categorical chain $..\langle \nu_{Pla} \rangle \triangleright \langle \nu_{Tre} \rangle \succ [\tau_{Rot}, \tau_{Trn}, \tau_{Stm}, \tau_{Lef}, \tau_{Flw}, \tau_{Frt}, \tau_{Sed}]$ is implemented as a set of categories and stored in a text format as follows:

"root CatPartTree CatPlantCatBiolObj;"
"leaf CatPartTree CatPlantCatBiolObj;"

The category of visual objects often refers to the specific knowledge domain and need to be learned in reference to this specific knowledge domain. For example, for the category of musical knowledge all categories of musical objects, described in [1] need to be learned. Learning of the category of the musical instruments is to learn all categories derived from the category of the musical instrument. The musical-instruments-category is divided into the percussion-category ν_{MIP}, the stringed-category ν_{MISt}, the keyboard-category ν_{MIK}, the wind-category ν_{MIW}, and the electronic-category ν_{MIE}, and is given by the following categorical chain: $\langle \Pi \rangle \triangleright \langle \sigma_{El} \rangle \triangleright \langle \nu_{ReO} \rangle \triangleright \langle \nu_{Ear} \rangle \triangleright \langle \nu_{NLiv} \rangle \triangleright \langle \nu_{MMad} \rangle \triangleright \langle \nu_{MIn} \rangle \triangleright \langle \nu_{MISt}, \nu_{MIW}, \nu_{MIP}, \nu_{MIK}, \nu_{MIE} \rangle$. Derived categories can be described as type of musical instruments. Examples of categories of stringed instruments are: the viola category ν_{Vio}, the cello category ν_{Cel}, the lute category ν_{Lut}, the balalaika category ν_{Bal} or the guitar category ν_{Git} and, using the categorical chain notation, are represented as follows: $.. \triangleright \langle \nu_{MIn} \rangle \triangleright \langle \nu_{MISt} \rangle \triangleright \langle \nu_{Vio}, \nu_{Cel}, \nu_{Lut}, \nu_{Git}, \nu_{Bal}, .. \rangle$. Similarly, a specific category derived from the guitar category is described as 'a type of guitar'.

Musical instrument	CatMusINSTR	
Stringed instrument	CatStringINSTR	CatMusINSTR
viola	CatStringINSTR	CatMusINSTR
guitar	CatStringINSTR	CatMusINSTR

Learning of the categorical chain is an important part of learning of the non-visual knowledge. During learning of the visual concepts of the selected ontological category the name of the category that is assigned to this category is also the name of one of the categories of the categorical chain. Learning of the names of the objects from a selected category is associated with learning of the knowledge of the categorical chain. The categorical chain is implemented as an array of interconnected names of the categories of the Categorical-Chain-Expert. During learning at the first stages of learning there is no need to supply names for all learned categories at a proper categorical level. During incremental categorical learning names of categories that were not learned can be added and inter-categorical links can be established.

Dependence Among Ontological Categories

The categorical chains can be linked and form a very complex structure of the knowledge called the categorical net that is used to interpret the visual object. The categorical net is used for the categorical reasoning that is based on 'moving' through the categorical net categories. In order to illustrate the dependence of the different ontological categories, the categorical net of the music objects is given as an example. The categorical net of the music objects link together the different visual categories such as the category of musical notation to provide the interpretational structure of the musical world (see Fig. 8.49).

Fig. 8.49. Example of members of the different visual musical categories

Musical notation is a visual record of heard or imagined musical sound, or a set of visual instructions for performance of music. It usually takes written or printed form. Musical notation serves as a means of preserving music over long periods of time, facilitates performance by others, and presents music in a form suitable for study and analysis. The categories of musical notation include the categories of musical symbols. The categories of musical notation supply knowledge that make it possible to understand the musical visual symbols, to play the musical composition and to record the composed musical work by writing it in the form of the musical scores. The category of the musical elements such as the note $\langle v_{Not} \rangle$, the rest $\langle v_{Res} \rangle$, or the clef $\langle v_{Cle} \rangle$, supply knowledge that makes it possible to recognize and name musical symbols. The categorical chain for the category of musical symbols is as follows: $\langle \Pi \rangle \triangleright \langle \sigma_{El} \rangle \triangleright \langle v_{Sg} \rangle \triangleright \langle v_{VSym} \rangle \triangleright \langle v_{Mus} \rangle \triangleright \langle v_{Not}, v_{Res}, v_{Cle}, .. \rangle$. The lower level of the category of musical symbols is the level of specific category of musical symbols such as the category of the bass clef, the treble clef or the C clef $.. \triangleright \langle v_{Mus} \rangle \triangleright \langle v_{Cle} \rangle \triangleright \langle v_{CTre}, v_{CBas}, v_{C\_C}, .. \rangle$. The knowledge of the specific category of musical elements makes it possible to interpret correctly musical symbols. Visual categories of musical symbols are linked with knowledge categories of musical symbols. The knowledge category of musical symbols supplies knowledge that makes it possible to interpret musical visual symbols as a specific musical sound $\kappa_{KB} \triangleright \langle \kappa_{KOb} \rangle \triangleright \langle \kappa_{Mus} \rangle \triangleright \langle \kappa_{Not}, \kappa_{Res}, \kappa_{Cle}, .. \rangle$. Understanding musical symbols means knowing how to make the sound using the musical instrument. The musical element such as notes refers to the elements of musical sound such as a pitch, or the location of musical sound on the musical scale.

The visual symbols used as the means of musical notation supply a set of elements that are used to form more complex musical expression. These complex expressions are formed according to the rules of the musical composition. For example, musical symbols placed on the staff are interpreted in terms of the musical notation and can be used to produce the musical sound. The musical symbols placed on the staff are members of the musical categories such as the category beaming, the category phrase, the rhythm category, the harmony category or the melody category (sequence of musical symbols on the staff) derived from the structural pattern categories given by the categorical chain: $\langle \Pi \rangle \triangleright \langle \sigma_{Pt} \rangle \triangleright \langle v_{Sg} \rangle \triangleright \langle v_{VSym} \rangle \triangleright \langle v_{Mus} \rangle \triangleright \langle v_{Bea}, v_{Phr}, v_{MLi}, .. \rangle$. The category of pattern of musical elements refers to interpretation of the sequence of musical symbols in terms of a melody or a rhythm. Sequence of musical symbols placed on the staff can be transformed into musical sound by musician playing on one of the musical instruments. Music is played by musicians who transform the music scores into musical sound. The category of musicians is derived from the category of professionals $\langle v_{Prf} \rangle$ and is given by the following categorical chain: $\langle \Pi \rangle \triangleright \langle \sigma_{El} \rangle \triangleright \langle v_{ReO} \rangle \triangleright \langle v_{Ear} \rangle \triangleright \langle v_{Liv} \rangle \triangleright \langle v_{Man} \rangle \triangleright \langle v_{Prf} \rangle \triangleright \langle v_{MCom}, v_{MMus}, .. \rangle$. There are two main categories of professionals that make music, the category of musicians v_{MMus} and the category of composers v_{MCom}. The musical works are composed by composers and can be performed by musician or a group of musicians (trio, quartet, and orchestra). Orchestra is instrumental ensemble of varying size and composition. The different musicians use the different instruments. The musicians use instruments to perform the musical work. The different musical instruments have to be used to play the different parts of the musical work. Musical instrument is any device for producing a musical sound. The principal types of such instruments, classified by the method of producing sound, are percussion, stringed, keyboard, wind, and electronic. Based on this classification of the musical instruments the category of the musical instrument is divided into the percussion category ν_{MIP}, the stringed category ν_{MISt}, the keyboard category ν_{MIK}, the wind category ν_{MIW}, and the electronic category ν_{MIE} and is given by the following categorical chain: $\langle \Pi \rangle \triangleright \langle \sigma_{El} \rangle \triangleright \langle \nu_{\mathrm{Re}O} \rangle \triangleright \langle \nu_{Ear} \rangle \triangleright \langle \nu_{NLiv} \rangle \triangleright \langle \nu_{MMad} \rangle \triangleright \langle \nu_{MIn} \rangle \triangleright \langle \nu_{MISt}, \nu_{MIW}, \nu_{MIP}, \nu_{MIK}, \nu_{MIE} \rangle$. For example, stringed instrument is any musical instrument that produces sound by the vibration of strings, which may be made of vegetable fiber, metal, animal gut, or plastic. In nearly all stringed instruments the sound of the vibrating string is amplified by the use of a resonating chamber or soundboard. The specific categories of stringed musical instrument are: the viola category ν_{Vio}, the cello category ν_{Cel}, the lute category ν_{Lut}, the balalaika category ν_{Bal} or the guitar category ν_{Git} and are represented as follows: $.. \triangleright \langle \nu_{MIn} \rangle \triangleright \langle \nu_{MISt} \rangle \triangleright \langle \nu_{Vio}, \nu_{Cel}, \nu_{Lut}, \nu_{Git}, \nu_{Bal}, .. \rangle$. The musical instrument is produced by instrument makers. Instrument makers specialize in production of the specific instruments such as a viola or a cello. The visual category of instrument makers ν_{MMIM} is derived from the category of professionals $\nu_{\mathrm{Pr}f}$. The visual chain for these categories is given as $.. \triangleright \langle \nu_{\mathrm{Pr}f} \rangle \triangleright \langle \nu_{MMIM} \rangle \triangleright \langle \nu_{IMVio}, \nu_{IMGit}, .. \rangle$. The music that is performed in the special building can be recorded and stored on the magnetic tape or CD. Music recording is physical record of a musical performance that can then be played back, or reproduced. Sound recording is transcription of vibrations in air that are perceptible as sound onto a storage medium, such as a phonograph disc. In sound reproduction the process is reversed so that the sound stored on the medium are converted back into sound waves. The three principal media that have been developed for sound recording and reproduction are the mechanical (phonographic disc), magnetic (audiotape), and optical (digital compact disc) systems. The recorded music can be played by using electronic devices such as gramophone or CD-player. The category of the electronic devices is derived from the pattern object to indicate that the devices are assembled from the simple elements. The electronic devices that are used for the purpose of music playing are members of the electronic sound device categories such as a gramophone, a radio, a TV-set, a tape-recorder or a CD-player. The electronic sound device is assembled from the electronic elements such as a transistor. The category of the electronic sound device is represented by the following visual chain $\langle \Pi \rangle \triangleright \langle \sigma_{El} \rangle \triangleright ... \triangleright \langle \nu_{MMad} \rangle \triangleright \langle \nu_{AsP} \rangle \triangleright \langle \nu_{EAsP} \rangle \triangleright \langle \nu_{\mathrm{Re}z}, \nu_{Ind}, \nu_{Tfo}, \nu_{Cap}, \nu_{Dio}, \nu_{Tran}, .. \rangle$.

This chain was described in [1], in the context of the description of assembling of electronic devices based on the electronic schema. Members of the category of electronic sound devices such as a gramophone or a CD-player are assembled based on the schema of electronic circuit. The category of the electronic sound devices is derived from the category of the electronic devices v_{ElDev} and is divided into the gramophone category v_{Gra}, the radio category v_{Rad}, the TV-set category v_{TV}, the tape-recorder category v_{Mag} or the CD-player category v_{CD}. The categorical chain of categories of the electronic sound devices is given in the following form: $\langle \Pi \rangle \triangleright \langle \sigma_{Pt} \rangle .. \triangleright \langle v_{Dev} \rangle \triangleright \langle v_{ElDev} \rangle \triangleright \langle v_{EDSo} \rangle \triangleright \langle v_{Rad}, v_{TV}, v_{Gra}, v_{Mag}, v_{CD}, . \rangle$.

The music is performed in the special places such as concert halls, opera houses, musical schools that are members of the musical house category. The musical house category derived from the house category is divided into the concert hall category v_{HCon}, the opera house category v_{HOper} or the musical school category v_{HSco} and is given by the following chain: $\langle \Pi \rangle \triangleright \langle \sigma_{El} \rangle \triangleright .. \triangleright \langle v_{MMad} \rangle \triangleright \langle v_{Hou} \rangle \triangleright \langle v_{HMus} \rangle \triangleright \langle v_{HOper}, v_{HCon}, v_{HSco}, . \rangle$. Opera houses are built by building workers and designed by architects that are members of the building worker category v_{Bul} or the architect category v_{Arh} represented by the following chain: $.. \triangleright \langle v_{Man} \rangle \triangleright \langle v_{Prf} \rangle \triangleright \langle v_{Arh}, v_{Bul}, . \rangle$.

Music is played by musician who uses the musical instrument to produce the music which is a special kind of the sound wave. The musical sound is characterized by physical properties of the sound wave. Sound results from the vibration of elastic bodies such as violin string or human vocal chords and is subject of research the branch of science called acoustic. Acoustic is the science concerned with the production, control, transmission, reception, and effects of sound. The sound can be visualized and considered as the sub-category of the category of the visual processes. The visualization of the sound wave can be obtained by applying the different transformations such as Fourier or Wavelets. The category of musical sound v_{Mus} is derived from the category of the acoustic processes v_{Acus} given by the following visual chain: $O \triangleright \langle v_{ReO} \rangle \triangleright \langle v_{Ear} \rangle \triangleright \langle v_{NLiv} \rangle \triangleright \langle v_{NatP} \rangle \triangleright \langle v_{Acus} \rangle \triangleright \langle v_{Mus}, v_{Son}, v_{Spi}, v_{Noi}, . \rangle$. The relations among categories are represented by the dependence diagram. The dependence diagram keeps the links to all visual categorical chains that are related to each other. Fig. 8.50 shows the dependence diagrams of categorical chains that are related to the music category. Based on the dependence diagram the visual objects such as the violin (the category of the musical instrument) can be interpreted in the context of the learned knowledge represented by linked categorical chains. The dependence diagram makes it possible to infer that musical instrument is used by musician to play the music that is composed by composer and that is given in the form of the musical scores. From the categorical chain of the category of the musical instrument we can have access to knowledge about the specific instruments such as a violin or a guitar. The dependence diagram makes it possible to establish any connection with all categorical chains of the dependence diagram and by this we can have access to knowledge supplied by both visual and knowledge chains. Each category of the dependence diagram have link to other dependence diagrams that has knowledge of the different aspects of the visual world. For example, the musical symbols that are part of the dependence diagram of the musical categories can give the link to the mathematical symbols that is part of the dependence diagram of the mathematical categories. Inference that is based on dependence diagrams is part of the thinking process that can offer nearly infinite possibilities of the creative exploration of the different categorical links.

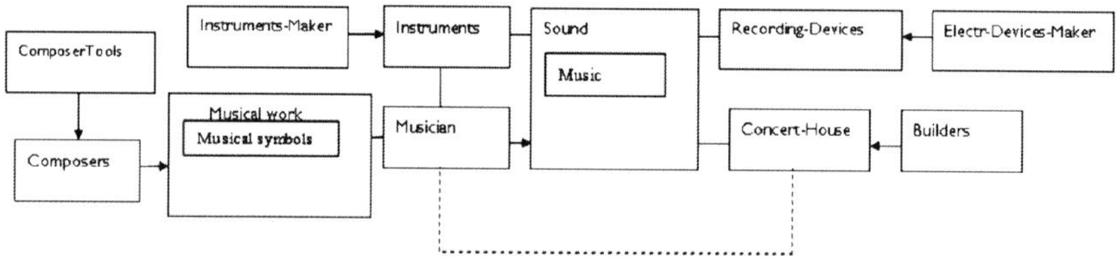

Fig. 8.50. The dependence diagram of the musical categories

8.2.1.2 The Knowledge Category

In this Section learning of the knowledge category is described. The knowledge category is the category that is derived from the categories of the selected scientific disciplines such as physics or chemistry. These categories are usually well defined and their meaning is established during scientific methodology of gathering of the scientific knowledge. The knowledge category can be properly interpreted only in the context of the categories of the same knowledge domain. For example, the velocity or distance in the domain of the kinematics as part of the physical domain is well defined and has a precise meaning. The knowledge category called the category of body of the knowledge $\left\langle \kappa_{BodK} \right\rangle$ is divided into the category of theology $\left\langle \kappa_{Teol} \right\rangle$, the category of philosophy $\left\langle \kappa_{Phil} \right\rangle$, the category of scientific discipline (the knowledge object) $\left\langle \kappa_{KOb} \right\rangle$ and the category of common sense knowledge $\left\langle \kappa_{KSK} \right\rangle$. The category of scientific discipline (the knowledge object)

$\langle \kappa_{KOb} \rangle$ is divided into the category of physical sciences, the biological sciences, the medicine, the engineering or the social sciences and can be denoted by knowledge chain as follows:

$$\kappa_{KB} \rhd \langle \kappa_{Teol} \rangle \rhd \langle \kappa_{Phil} \rangle \rhd \langle \kappa_{KOb} \rangle \rhd \langle \kappa_{PhS}, \kappa_{BiS}, \kappa_{MeS}, \kappa_{EnS}, \kappa_{SoS}, . \rangle .$$

The knowledge categories are usually well defined within the range of a given scientific discipline. For example, the mineralogy as a part of the earth science has its own well defined categories such as the mineral category, the chemical composition, the crystalline structure or the rock category. The mineral category or rock category is related to the categories of the visual object such as the mineral category or the rock category; however meaning of these categories is different. Coding categories that are derived from the knowledge object such as the physical science are used during learning of the query-form $\Theta(T)$ of the physical text-task. The coding categories such as [CObjMov CActMove CNum [CUv] CAuxIn CNum [CUt] CQCalcul CatPhisObjSpd] are described in the following Sections.

8.2.1.3 The Coding Category

In this Section learning of the coding categories will be presented. The coding category is a category that is established during the iterative learning process and usually is derived from the ontological categories of the visual object v_i or knowledge categories κ_i . The coding categories derived from the category of visual objects (described in [1]) are rather well established. These categories are usually part of the knowledge of the specific domain. For example, the coding categories derived from the figure category such as polygons or quadrilaterals are part of the knowledge of geometry.

The Coding-Category-Expert implements the coding categories as a searching procedure and data stored in the text format in categorexicon.txt file. For example, the categories derived from the category of the visual object represented by the following categorical chain: $O \rhd \langle v_{Fig} \rangle \rhd \langle v_{2DF} \rangle \rhd \langle v_{Pol} \rangle \rhd \langle v_{Qud} \rangle \rhd \langle v_{Rec} \rangle$ are stored in the categorexicon as follows:.

rectangle	CRect	CQuad	CPolygon
	C2DDFig	CtFigure	

Coding categories are used to define the query-form. During learning of the query-form the categories of the visual object can be selected at any categorical level. For example, the word a "rectangle" can be coded as Crect, CQuad, CPoly, or C2DFig. This way of coding makes it possible to find the optimal level of the query-form that can cover the broad range of the meaning of the text and at the same time do not lead to confusion by mixing texts having different meaning.

During SUS learning, the names of the visual categories such as CatRect, CatPolygon, CatQuad, Cat2DFig, CatFig are often learned as names of the coding categories such as CRect, CPoly, CQuad, C2DFig, CFig, by changing part "Cat" into "C'. The Coding-Category-Expert is invoked during learning stage to acquire data that are stored in the categorexicon. The Coding-Category-Expert invokes Excel program to facilitate inserting the data, provide method of a provisional data checking, and stores the categorexicon in a text format.

The category of visual objects can be associated with one or more than one coding categories. For example, the word an "apple" that is a member of the category of visual objects can be associated with the coding categories "CFruit", "CEatObj", or "CMarketObj", that means, it is considered as an object of the botanical science, an ingredient of the food, an object that can be sold on the market:

apple	CFruit	CBotObj	CBiolObj
apple	CFruitFood	CBotFood	CFood
apple	CFruitMarket	CMarketOBJ	CMarket

In the case when contextual information is accessible, a text is related to the given domain e.g. physics, the coding categories are part of the categorexicon of the given domain e.g categorixicon of physics PhisCategorixicon. For example, in the case when the coding category is related to the domain of physic the word an "apple" is associated with the coding category moving object "CMovObj".

The specific coding category derived from the category of visual objects can be specified by values of the one attribute or more than one attributes and represented by the text-queries $[C_i](p_i)$. A name of a specific coding category can be a new name or can consists of both, the name of the general category and the name of the feature as an adjective. For example, Jonathan is the new name for the specific apple category derived from the apple category having value of its attributes different from other apples described by the different specific categories. The compound names for the specific coding categories, derived from the nail category such as "ring nail" or "wire nail", are examples of names of the specific categories given by two words, the name of the value of predicate and the name of the general category. A ring nail is a nail with annular rings on its shaft, whereas a wire nail is a nail that has very small head.

An important derivation mechanism is given by the feature values that are specified by the marker such as greater, small or higher. The compound name of the derived category is usually combination of the name of the marker and the name of the general category and is often the synonym of the specific name of the derived category. For example, the highest peak is the synonym for Mount Everest. Learning of categories that are synonymies and consist only one specific category is like learning of the synonym. In the case when the compound name is referring to more than one category names of these categories can be stored or obtained by invoking the Compound-Name-Expert. For example, the coding categories derived from the specific category given by the name 'small animal' can be obtained by sorting data in the "animal category" based on the size attribute and the n last cases are selected to be members of the category "small animal".

animal	size
rabbitt	small
elephant	large
mouse	small
dog	medium

Or

animal	size
rabbitt	20
elephant	200
mouse	5
dog	80

Sort size

animal	size
elephant	200
dog	80
rabbitt	20
mouse	5

The coding categories are divided into the object coding category c_i^V , the action coding category c_i^A , and the attribute coding category c_i^F . During iterative learning the coding categories are learned in the context of the selected knowledge domain and are based on the accessible knowledge of this domain. When the knowledge categories K_i from the knowledge domain such as physics or mathematics are well defined categories of common sense knowledge often do not have well defined meaning. One domain of common sense knowledge is knowledge of the tools category. The tools category CTool is divided into the fastener categories CFASTEN={screw, nail}, the category of blowing tools CBLOWTOOL={chisel, ..}, or the category of material CMATERIAL={stone, brick, …}. The action tool category is divided into category of different activities such as the category of tool action CActTOOL={cut, saw, blow, drive, strike, flattening, tap …}, the category of removing CActREMOVE={extract, remove, pulling out, eject }, the fitting category CActFIT={fit, ..}, or the blowing category CActBLOW={blow, strike}. The attribute tool category is divided into the geometrical category CAtrGEOMETRY={annular, circular, elliptical, ..} or the tool attribute CAtrTOOL={ small narrow, small flat, very small, stubby, compressed, diamond-shaped, wide, large}.

The coding categories are stored in the text format as the categorexicon file that is loaded by the Coding-Category-Expert. The General-Categorexicon-Expert is invoked during understanding of the unknown text (unknown domain). In the case when there is an information concerning the knowledge domain the specific categorexicon expert is invoked. For example, when the physic-text-task is to be solved the Physical-Categorexicon-Expert is invoked. The Physical-Categorexicon-Expert loads the physical categorexicon that includes predefined coding categories that have a proper meaning in the context of the physical text. In this categorexicon, an apple is coded as a CMovObj without reference to other properties of this object. In the explanatory process the Explanatory-Script-Expert can refers to the General-Categorexicon-Expert to explain some contextual information.

8.2.2 Implementation of the New Knowledge – The Knowledge Schema

In this section learning of the knowledge schema is presented. A knowledge schema contains the basic knowledge needed to interpret the perceived object. For each learned category of the visual knowledge the knowledge schema is learned as part of the knowledge of this category. The knowledge schema includes the name of the category ∂_{Nam}, the visual concept ∂_{ViC}, the definition ∂_{Def} and the method of the exemplar generation ∂_{MGe}, and is denoted as: $..\langle \kappa_C \rangle \prec \{\partial_{ViC}, \partial_{Nam}, \partial_{MIn}, \partial_{Def}, \partial_{MGe}\}$. The name ∂_{Nam} is given in the form of a linguistic description expressed in one of the existing languages and is denoted as follows: $..\langle \kappa_C \rangle \prec \{\partial_{Nam}\} \| v_{NLan} \| \langle v_{C1}, v_{C2}, ..., v_{CN} \rangle$, where $\langle v_{C1}, v_{C2}, ..., v_{CN} \rangle$ represents the categories of names that depend on the selected language category $\| v_{NLan} \|$. The definition ∂_{Def} gives a short description of the visual object. The visual concept for each learned category is given as a set of symbolic names $\varphi_c^1, ..., \varphi_c^K$, where K is a number of learned categories. The method of the exemplar generation ∂_{MGe} specifies an algorithm of exemplar generation.

The knowledge schema is implemented as the Knowledge-Schema-Expert that invokes the Naming-Expert, the Visual-Concept-Expert, the Definition-Expert and the Exemplar-Generating-Expert. During learning process the Naming-Expert is invoked to learn the name in English as well as in other languages and synonyms in English and other languages. The Visual-Concept-Expert is invoked to learn visual concepts. Learning of the visual concepts was presented in previous chapters. The Definition-Expert is invoked to learn definition in the form of the short description. The Exemplar-Generating-Expert implements the different algorithms for exemplar generation. The Knowledge-Schema-Expert invokes the Naming-Expert and the Visual-Concept-Expert to search for the name of the category of the examined object during naming process. During naming process, an examined object u is transformed into the symbolic name η and next the Visual-Concept-Expert is searching for a visual concept that matches the symbolic name η. When the visual concept φ_h is found the name of the category c_h is used as the name of an examined object.

The symbolic name η is used to generate an exemplar e from the specified class (given by the selected symbolic name) by invoking the Exemplar-Generating-Expert. The Exemplar-Generating-Expert invokes an appropriate Model-Expert to specify the parameters of the model. Generating of the complex object requires using many parameters given in the form of the symbolic description κ_k. The symbolic description κ_k as well as the symbolic name η_k (SUS notation) are stored in a text format and are processed by the Symbolic-Name-Expert. The Symbolic-Name-Expert reads the symbolic names from the text file and stores these symbolic names as an array of strings.

The visual knowledge in the form of the visual concept φ_h of the specific category is learned at the prototype level. The prototype is the category at the bottom of each categorical chain. The prototype is defined during learning process at the level for which the training exemplars are available. The prototype is represented by all visual representatives of a specific category and by this the visual knowledge is covering a visual domain of the prototype. The visual domain prototype refers to the visual knowledge that makes it possible to recognize all visual representatives of the prototype.

8.2.3 Implementation of the New Knowledge – The Text-Query Category

In this Section learning of the knowledge of the text-query is described. As it was described in the previous chapters, the text-query T^G is the text that does not have a query-part $T^G \equiv T_U^T$. The text-queries are often used in crossword puzzles and are implemented in the context of the Crossword-Puzzle-Expert. Crossword puzzle is a puzzle in which words crossing one another, usually horizontally and vertically in a chequered pattern of squares, has to be filled in from clues provided. The synonym-text-queries $T^G[S]$ are often used in crossword puzzles. The synonym-text-queries $T^G[S]$ can be easily transformed into the text-task $T^T[S]$ by adding the query-part $T^T[S] \equiv T_U^T T_Q^T[S]$. For example, the synonym-text-query "wheelie bin" can be transformed into the text-task "another name for wheelie bin". The Crossword-Puzzle-Expert invokes the Synonym-Expert to learn and to understand synonymies. It should be noticed, that the range of meaning of the synonymies used in our research not always covers a range of meanings of a given synonym that can be found in dictionaries. The range of meanings of the synonym depends on the domain to which this synonym belongs. Learning of the synonymies requires learning all synonymies for all learned names. Learning synonymies of categories of visual objects is related to the knowledge domain of those categories. For example, the synonym for mineral's name "chiastolite" is a list of names [Chyastolith, Cross-stone, Crucite, Crusite, Lapis Crucifer, Macle, Maltesite]. Learned synonym is given as both a name and a synonym that is marked as the synonym category CSynonym. For example, crucite is the synonym of chiastolite. The learned synonyms are stored in a text file synorexicon in the text form as follows:

[Crucite, >>CSynonym, >>chiastolite, >>CMineral].

The synonym is learned for each ontological category by invoking the Synonym-Expert. During understanding process (e.g. during solving a crossword puzzle) the Crossword-Puzzle-Expert invokes the Synonym-Expert to find synonymies for a given name. The Crossword-Puzzle-Expert implements different methods of solving crossword puzzles. A synonyms method of solving crossword puzzles is a method based on finding a list of synonyms and next searching for the name that fulfils a specific criterion. The specific criterion specifies a number of letters of a synonym (word) and the matching algorithm is used to match the letters that are previously ‘guessed’. For example, let’s assume that the question is given as follows: (*capsule*)? and the name consisting of four letters (four ‘cells’ $\boxed{\ }\boxed{\ }\boxed{\ }\boxed{\ }$ need to be fulfilled). A list of synonymies for the *capsule* is as follows: capsule [pod, container, case, casing, shell, pill, tablet]. Searching through the list [pod, container, case, casing, shell, pill, tablet] the following possible categories are selected: [case, pill]. In order to select one word there is a need to find an additional letter in the cell. Lets assume that the letter “a” is found by solving another part of the crossword puzzle $\boxed{\ }\boxed{a}\boxed{\ }\boxed{\ }$. As a result of filling pattern the following solution is found: ‘case’. The Crossword-Puzzle-Expert at first, selects all possible solutions for each pattern of squares and at the end the final solution is obtained by matching the possible patterns.

The categorical level text-query $T^G[K]$ has often a very concise form and refers to the different categorical levels of different categories. For example, the text-query “a general term for beef, lamb, venison, etc.” refers to the general category “meet” from which the specific categories such as ->[beef, lamb, venison, …] are selected. Such question can be answered by searching of the categorical chain $\ldots v_{Ani} \triangleright v_{Fod} \triangleright v_{Met} \triangleright \left\langle v_{Bef}, v_{Lam}, v_{Ven}, \ldots \right\rangle$ meet->[beef, lamb, venison, …]. During solving the crossword puzzle with the general term text-queries the Crossword-Puzzle-Expert invokes the Categorical-Chain-Expert to solve the problem. The Categorical-Chain-Expert described in the previous chapters, implements the method of searching for the category of more general or more specific level of the categorical chain.

The type-text-query $T^G[typ](C_i)$ is learned by transforming it into the query-form $\Phi(T_Q^G[typ])C_i))$ “typeObj” e.g. the type-text-query “type of hammer” is transformed into the query-form “typeHandTool” or “typeHammer”. Knowledge for the query-form of the type text-query $T^G[typ](C_i)$ is stored in the form: [“Hammer, CtypeHammer, mallet, gavel, maul”] in a text format in the typecategorexicon.txt file.

During solving the crossword puzzle with the type-text-queries, the Crossword-Puzzle-Expert invokes the Type-Expert to solve the problem. The Type-Expert implements searching algorithms to search typecategorexicon.txt file.

Examples of learned type-text-queries: “type of clarinet"; “type of nut"; type of vegetable"; “type of apple"; “type of fruit"; “type of flower"; “type of tree"; “type of palm"; “type of dog"; “type of insect"; “type of reptile"; “type of earthworm"; type of invertebrate"; “type of animal"; “type of phylum"; “type of bean"; “type of vegetables"; "type of seed"; “type of veal"; “type of goose"; “type of celery"; “type of fat"; “type of condiment"; “type of herb"; “type of fowl"; “type of fish"; “type of seafood"; “type of yogurt"; “type of fruit"; type of sorghum"; “type of millet"; "type of barley"; "type of berry"; "type of dowel"; "type of washers"; "type of nails"; type of fasteners"; type of wrench"; "type of vice"; "type of trowel"; "type of scissors"; "type of pliers"; "type of saw"; "type of mallet"; "type of tools"; type of lever"; "type of knife"; type of spoon"; "type of microscope"; "type of wind instrument"; "type of prehistoric reptile"; "type of natural sweetener"; "type of artificial flavoring"; "type of hot chisels"; "type of woodworking chisel"; "type of gardening scissors"; type of circular saw"; "type of metalworking hammer"; "type of driving tools"; "type of hand tools"; "type of defining tools"; "type of mechanically-powered hammer"; "type of simple machine"; "type of instrument for measuring electricity", "type of measuring instrument used in psychology", "brad of horse";

Categories similar to the type category are represented by words such as "cut, breed, instrument, method, processing, operation, process, constituent, form, property, product, ingredient". Example of those categories used in learning of the text-query: "herb mixture"; "beef cut"; "duck breed", "cooking instrument"; "cooking method"; "sherry processing"; "food process"; "fruit operation"; "milk constituent"; "chocolate form"; "bakery ingredient". These text-queries were tested by transforming text-query such as "type of knife" into the text-task "give name of knife type" or "list all names of the knife types".

The part text-queries $T^G[part](C_i)$ are learned by transforming part-text-queries into the query-form $\Phi(T_Q^G[part])C_i$)) "partObj". The text-query that refers to the part category is often described by the term "structure, morphology, anatomy, section, composition, system, organs, elements, for example, "moss structure", "spider morphology", "snail anatomy", "poppy capsule section", "digestive system", "sense organs", "elements of a house". The knowledge for query-form of the part-text-query $T^G[part](C_i)$ is stored in the text format as partcategorexicon. During solving the crossword puzzle with part-text-queries the Crossword-Puzzle-Expert invokes the Part-Expert that implements searching algorithms to search partcategorexicon.

The attribute text-queries $T^G[C_i](p)$ that has one attribute, two attributes $T^G[C_i](p_1, p_2)$, three attributes $T^G[C_i](p_1, p_2, p_3)$ or more than tree attributes $T^G[C_i](p_1,..., p_N)$ are learned by transforming it into the query-form $\Phi(T^G[C_i](p_1,..., p_N))$. The knowledge of the query-form of the attribute text-query $T^G[C_i](p_1,..., p_N)$ is stored in the text format as attributecategorexicon. During solving the crossword puzzle with attribute text-queries the Crossword-Puzzle-Expert invokes the Attribute-Expert to solve the problem. The Attribute-Expert implements searching algorithms to search attributecategorexicon. The qualitative attributes such as size has its value given as the set of categories $val(p) = \{c^1,...,c^N\}$. The new category is specified by the selection of the one of the category such as small. Selecting the attribute category c^i "small" for the category animal will result in derivation of the new category $[C_i](val(p_i)=c')$ small animal. In the case of the text-query "small animal" the category C_i is "animal" and the value of attribute size p_{siz} is "small". Similarly the categories can be derived by specifying two or three attributes. Examples of learned attribute text-queries for categories derived from the visual object specified by values of the two attributes $T^G[C_i](p_1, p_2)$: "American wild cat", "long-horned NZ insect", "great US lake", "red pepper spice", "small thin sausage". Examples of categories derived from other categories specified by two features: "world wide web", "computer chip maker "international sporting event", "aboriginal musical instrument", "honey-based alcoholic drink". Examples of categories derived from the visual object specified by values of three attributes $T_N^G[C_i](p_1, p_2, p_3)$ are: "fast swimming oceanic fish", or "soft white Italian cheese".

The definition text-queries $T^G[Def]$ of the tools domain are learned based on the examples of the different forms of the definitions. The Definition-Expert invokes the Learning-Expert that stores the basic-form and the query-form in the text format. During understanding process (e.g. solving the crossword puzzle) the Definition-Expert invokes the Query-Expert, Basic-Expert and the Procedural-Expert to search for solution.

For the consist-text-query $T^G[\Im^{Cons}]$ the basic-form $OthatConsist[P_1,...,P_N]$ and the query-form "CTool CConsist CPartTool", where coding category CConsist includes marker words CConsist={that consist of, that has, with}, are learned. For the made-text-query $T^G[\Im^{Made}]$ the basic-form $Omade[M]$ (o is O that is made of material M) and the query-form "CTool CMade CMaterTool" are learned. For the used-material text-query $T^G[\Im^{Use}(M)]$ the basic-form $Oused[A \triangleright M]$ (o is O that is used for performing action A on material m) and the query-form "CTool CUsed CActTool CMaterial", where the coding category CUsed={for, used for, that is used for}, are learned. For the used-result text-query $T^G[\Im^{Use}(R)]$ the basic-form $Oused[A \triangleright R]$ (o is O that is used for performing action A to obtain result R) and the query-form "CTool CUsed CActTool CResult".where the coding category CUsed={for, used for, that is used for} are learned. For the used-as text-query $T^G[\Im^{UseAs}]$ the basic-form $OusedAs[T]$ (object o is object O used as tools) and the query-form "CTool CUsedAs CMTool" are learned.

In order to learn knowledge from a given domain there is a need to learn the definition that describe the different aspects of the defined object. The variation of the description can be learned as the specification of the general form. For example, the definition "nail used to attach wooden parts into concrete" can be given as "nail used to attach parts into concrete", "nail used to attach wooden parts", "nail used to attach wooden parts into material" or "nail used to attach different parts into material". The complex definitions can be transformed into the simple form by omitting some categories or some aspects of the definition, for example "nail that is similar to common nail but has slimmer shank and is used on lighter pieces of wood and on boxes" can be given in simplest form as "nail that is similar to common nail but has slimmer shank", "nail that is similar to common nail and is used on lighter pieces of wood and on boxes", "nail that is similar to common nail but has slimmer shank and is used on lighter pieces of wood", "nail that has slimmer shank and is used on lighter pieces of wood and on boxes", or "nail that is used on lighter pieces of wood and on boxes".

The text-queries that belong to the different knowledge domains are learned based on the selected queries of the crossword puzzles. Examples of the learned queries from the different domain used for learning text-queries: history "along with Callicrates who designed Parthenon"; film "Polish filmmaker who directed three colours trilogy"; politics "last prime minister of British Guiana"; "Russian political leader".

8.2.4 Implementation of the New Knowledge – The Text-Task Category

In this Section learning of the knowledge of the text-task is described. A text-category refers to any written script or text. For the purpose of this research, the text-category T is divided into four different specific categories: the category of text-query T^G, the category of text-task T^T, the category of dictionary-text T^D, and the category of long-text T^L. The text is interpreted in terms of its meaning. The meaning for the dictionary-text T^D is usually given by the script $\beta(T^D) \triangleright S$, whereas the meaning for the text-query and the task-text is given by both the explanatory script and the basic-form: $\beta(T^G) \triangleright (J,B)$ and $\beta(T^T) \triangleright (J,B)$. The text-task have both the task part T_Q^T and query-part that is denoted as $T^T = T_U^T T_Q^T$. The query-part T_Q^T is a sentence or a part of sentence that has a query term such as "what" or "who". The text-task has always the query part. The text-query T^G can be represented by one word or more than one word and does not have the query part.

Understanding of the dictionary-text T^D requires that SUS understands the contents of the text, is able to explain the meaning of the text and to answer questions that refer to the meaning of the text. Understanding of the dictionary-text is based on the previously learned interpretational script. The script includes the description of the stereotypical situation and is given at the different levels of description that reveals the different levels of details needed for understanding of the dictionary-text.

A task that describes a real world scene is called the text-task and is denoted as T^T. The text-task T^T can have different forms, can consist of different categories and can refer to different phenomena. The text-task can be given in the form of query or short description. The meaning of the text-task that refers to finding of the solution and is given by the basic-form, is called the basic meaning $\beta^B(T^T) \triangleright B$, whereas the meaning that refers to the interpretation of text in terms of the real world categories and is given by the interpretational script J, is called the interpretational meaning $\beta^J(T^T) \triangleright J$.

The text-task T^T is given in the form of questions, problems or tasks that can be found in the school textbooks, school tests, IQ tests, university handbooks, or crossword puzzles. In this book learning of the category of the educational-text-task such as the mathematical-text-task $T^T[E(Mat)]$, the physical-text-task $T^T[E(Phys)]$ or the geographical-text-task $T^T[E(Geo)]$ is presented. The text-task $T^T = T_U^T T_Q^T$ consists of two parts the query-part T_Q^T and the task part T_U^T and the query-form $\Theta(T^T)$ can be learnt independently as the query-form of the query-part $\Theta(T_Q^T)$ and the query-form of the task-part $\Theta(T_U^T)$. Query-forms are composed of the coding categories of the selected specific knowledge domains such as physics, geography or mathematics. The important part of the learning of the query-form is generalization based on the selected examples of the text-task. Learning of the text-task is an iterative process. At each stage of learning new examples of text-task are tested in the context of the previously learned knowledge.

The text-task T^T consist of the text-part T_U^T and the query-part T_Q^T. The text-part T_U^T can be often given in the form of the definition $\Im$. The text-task can be obtained by combining the query-part with the text-part $T_1^T = T_U^{T1} + T_Q^{T1} = T_U^{T1} T_Q^{T1}$. The explanatory query-part "what does it mean" can be combined with definition text-part $T_{EX\_DEF}^T = T_{U\_DEF}^T + T_{Q\_EX}^T$, for example, "what does it mean small sharpened metal object" is combination of the query-part "what does it mean" and text definition "small sharpened metal object". The query-form of the query-part is given as Q_whatMean, Q_WhatIsName. The ability to interpret precisely the meaning of the text depends on knowledge gathered during learning stage. In order to interpret precisely the meaning of the text the structural variation of the text need to be learned. However, in order to solve the task only the key categories need to be recognized so the learning process can be significantly improved.

The most difficult problem in learning of the text-task is to cope with the varieties of forms that have equal or nearly equal meaning. Understanding requires interpreting the text to identify the content and to be able to find the difference among meaning of the text. The meaning of the text-task depends on both the interpretational meaning and the basic meaning. As it was shown the two text-tasks have the same basic meaning if both refer to the same basic-form $\beta^B(T_{U1}^T T_{Q1}^T) \equiv \beta^B(T_{U2}^T T_{Q2}^T) \triangleright B$. Similarly the two text-tasks that differ in query-part $\beta^B(T_U^T T_{Q1}^T) \equiv \beta^B(T_U^T T_{Q2}^T) \triangleright B$ or text-part $\beta^B(T_{U1}^T T_Q^T) \equiv \beta^B(T_{U2}^T T_Q^T) \triangleright B$ and refer to the same basic-form B have the same basic meaning $\beta^B(T^T)$. Learning of the text-task requires to learn the query-form $\Theta(T^T)$, the basic-form B, the procedural-form $P(B)$ and the explanatory script J. Understanding of the text-task depends on the domain knowledge to which the text-task belongs. In the following chapters learning of the mathematical-text-task $T^T[E(Mat)]$, the physical-text-task $T^T[E(Phys)]$ and the geographical-text-task $T^T[E(Geo)]$ is presented.

8.2.4.1 The Text-Task Category – The Query-Form

In this section learning of the query-form is presented. The query-form $\Theta(T^T)$ is the generalization of the text-task expressed in the form of the clauses consisting of the coding categories. Query-forms are composed of the coding categories of the selected specific knowledge domain such as physics, geography or mathematics. The important part of the learning of the query-form is generalization based on the selected examples of the text-task.

At the first stage of learning, one of the categories of the educational-text-task such as the mathematical-text-task $T^T[E(Mat)]$ is selected. At first, the-text-tasks used for learning are grouped into groups that have the same basic meaning and each text-task T_i^T is transformed into the basic-form $\beta_j(T_i^T) = B_j$. Next, the text-tasks are grouped to groups that have the same basic-form $T_i^T \in T_i^T(B_j) if \beta_j(T_i^T) = B_j$. These groups are arranged according to the difficulty level. Each group of tasks $T_i^T(B_j)$ is further grouped into groups that have the same query-form $T_i^T \in T_i^T(B_j(\theta_k)) if \Theta_k(T_i^T(\beta_j)) = \theta_k$ and the query-form $\Theta_k(T_i^T)$ is learned for each group of tasks $T_i^T(B_j(\theta_k))$. For each group of tasks $T_i^T(B_j(\theta_k))$ the query-form $\Theta_k(T_i^T)$ is learned by transforming text-tasks into a set of coding categories $\Theta_k(T_i^T) \equiv \theta_k = \{c_k^1, ..., c_k^N\}$, where N is a number of coding categories k-th query-form θ_k. For the text-task T_i^T, the word of the text $w_i \in T_i^T$ is coded by applying the coding categories c_i stored in the categorexicon $c_i \in C$. As the result of learning, the query-forms that consist of the coding categories $\theta_k = \{c_k^1, ..., c_k^N\}$ are obtained.

During understanding process the learned query-forms θ_k are used to find the basic-form, the procedural-form and the explanatory script. During understanding process an examined text-task T^0 is transformed into the set of coding categories $t = \{k^1,...,k^M\}$, where M is a number of coding categories. The first level of understanding process can be described by following algorithms:

For k=1 to K if $t \equiv \theta_k$ than apply $[B_k, P_k, J_k]$, END.

In the case when forms $[B_k, P_k, J_k]$ are not found that means that SUS can not solve the problem.

The learned query-forms θ_k are implemented as a part of the Query-Form-Expert. The query-form θ_k is implemented as the clause $if((k_i^1 \in \theta_k) \wedge (k_i^2 \in \theta_k),...,\wedge(k_i^M \in \theta_k))thanE(B_k)$, where $E[B_k]$ denotes the basic-form. If the clause is fulfilled the Basic-Form-Expert which implements the basic-form B_k is invoked.

The text-task $T^T = T_U^T T_Q^T$ consists of two parts, the query-part T_Q^T and the task part T_U^T. The query-form $\Theta(T^T)$ can be learned independently as the query-form of the query-part $\Theta(T_Q^T)$ and the query-form of the task part $\Theta(T_U^T)$. The generalization of the query-parts is obtained by analyzing a big number of examples. At first the preliminary query parts are obtained and next the coding categories are used to obtain the query-form. For example, for the text-task, "Student purchase four object for $ 2 each what is total cost", the query-part is "what is total cost" and can be exchanged into the query-part"compute total cost", "find cost of all objects", "calculate cost of four pens", "compute cost", "compute cost of all bought objects", CQuery CBe CAll CCost ={"what is total cost"}, CCompute CAll CCost={"compute total cost"}, CFind CAll CCost CObj ={find cost of all objects}, CFind CCost CNum CObj ={find cost of 4 objects, find cost of four pens }. By inserting the coding category such as CQ_Compute we can write the query-part in the general form e.g. "CQ_Compute total cost" or "CQ_Compute cost of all objects". In similar way we can establish the coding category CCost={total cost, cost of all objects, cost of N objects, cost} so the query-part can be expressed in general form as CQ_Compute CCost. So the query-form for the text-task "Student purchase four object for $ 2 each what is total cost", is CMAN CBuy N CObj for N cU CQ_Compute CCost. This general form can be used to translate the similar text-task into the basic-form.

The clauses for the query-form $\Theta(T^T)$ divided into the query-part Q and the task part U are implemented in the form $if(Q \in \theta_Q)if(U \in \theta_U)thanE[B_k]$.

The Query-Form-Expert implements the query-form as the sequence of the clauses $if(Q \in \theta_Q)if(U \in \theta_U)thanE[B_k]$. The knowledge of the specific knowledge domain such as mathematics is implemented by the Mathematical-Query-Form-Expert. For example, for the text-task type "Kate buys 4 apples for $ 12 each what was totat cost" the query-form is given as follows:

```
if(m_MulFlagQuest>0 && m_MulFlagVerbBe>0 && m_MulFlagPartlCatTotal>0
&& m_MulFlagCatCost>0)//what was total cost
{
if(m_MulFlagCatName>0 && m_MulFlagCatBuy>0 &&
m_MulFlagBiolObjFruit>0)//Kate bought 4 apples
{
if(m_MulFlagFor>0 && m_MulFlagCurrencySymbolCurrencyCat>0 &&
m_MulFlagPronEach>0)//for $ 12 each
{
A_buy_n_o_for_c_each_Find_total_cost();
}
}
```

Each clause is build from the unified coding categories that are marked as Flag and has form if(A>0 && B>0 ....) than BasicForm_n(). The unification is aimed at obtaining better generalization of the query-form. The method BasicForm_n() invoke the Mathematical-Basic-Form-Expert. The method A_buy_n_o_for_c_each_Find_total_cost() invoke the Mathematical-Basic-Form-Expert. The clauses are divided into meaningful segments that can be easily exchanged to obtain the more specific or more general form. In this example, the text is divided into three segments: the query-part "what was total cost" and two segments "Kate bought 4 apples" and "for $ 12 each".

The system "Domeyko" that is used for understanding of the mineralogical dictionary-text use the coding categories CAT_ and method COMPARE() to find the value of the m_NoTruth parameter. If conditions (m_NoTruth>=m_NoZdanie) is fulfilled the Mineralogical-Basic-Form-Expert is invoked by Mineral Knowledge-Expert "pMineralKnowledgeExpert->It_Solve_In_Fuse(&m_INPUT);" Example of the implementation of the query-form for the dictionary-text "Cobaltite is soluble in nitric acid and is fusible" is as follows:

```
m_StringZDANIE[0]="CAT_MINNAME";
m_StringZDANIE[1]="CAT_PROPERTYSOLVE";
m_StringZDANIE[2]="CAT_IN";
m_StringZDANIE[3]="CAT_PROPERTYFUSE";
m_NoZdanie=4;m_NoTruth=0;
COMPARE();
if(m_NoTruth>=m_NoZdanie){pMineralKnowledgeObject-
>It_Solve_In_Fuse(&m_INPUT);}
```

Example of learning of the query-form for physical-text-task and mathematical text-task are given in the following chapters.

8.2.4.2 The Text-Task Category – The Basic-Form

In this section learning of the basic-form is presented. The basic-form refers to the basic meaning of the text-task. The meaning for the text-query and the text-task is given by both the script and the basic-form $\beta(T^G) \triangleright (J,B)$ and $\beta(T^T) \triangleright (J,B)$, and the basic-form contains only information needed to find the solution. The two text-tasks have the same basic meaning if both refer to the same basic-form $\beta^B(T_1^T) \equiv \beta^B(T_2^T) \triangleright B$.

The basic-form of the text-query refers to the definition. The important definitions are definitions that follow the schema that is independent of the knowledge domain such as the type definition. The type definition $\aleph_m^{Uf}$ can be translated into the basic-form by using the terms "for", "used for", or "that is used for". The type definition $\Im_m^{Uf} \equiv \{o\} \triangleright [O](UsedFor)\{A\}(m)$ means (object o is object O used for performing action A on material m) and is represented in the basic-form as "O is o used for ActCat Mat", for example the definition (saw is a tool that is used for cutting wood) or (hammer is a tool that is used for breaking stone). The type definition $\Im_R^{Uf} \equiv \{o\} \triangleright [O](UsedFor)\{A\}(R)$ means (object o is object O used for performing action A to obtain result R) and is represented in the basic-form as "O is o used for ActCat Res", egg. (die tools for cutting thread).

The definition type $\Im_R^{Uf}$ ("used for") can be exchanged by the definition type of $\Im_R^f$ ("for") for example "saw is (tool) for cutting (wood)" ->"saw is (tool) used for cutting (wood)" or "saw is (tool) that is used for cutting (wood)".

The type definition $\Im_R^{Ut} \equiv \{o\} \triangleright [O](UsedFor)\{A\}(R_T)$ means (object o is object O used to performing action A to have result R_T) is represented by basic-form as "O is o used to ActCat ActTool" e.g. "hammer is used to deliver a blow".

The type definition $\Im_T^{Ua} \equiv \{o\} \triangleright [O](UsedAs)(T\ )$ means (object o is object O used as tools) is represented in basic-form "O is o used as CatTool" e.g. "crowbar is a tool used as a lever".

Implementing of the query-form from the definition given in the form of specification of the values of attributes of the object (category) $\Im^p\{o\} \equiv O(p)$ can be given in the general form CAtr CObj. However this query-form has categories that are too general and can cause the misinterpretation problem. The more specific query-form is strictly connected with the one of the domain knowledge to which a given definition belongs. For example for the tool domain the query-form is given as CAtrTool CObjTool. The CAtrTool is the specific query-form in the tool domain where the category CAtrTool and CObj include subspecific categories that can also belong to other knowledge domains.

The definition that is part of geometrical knowledge can be given in terms of its parts $\Im^{P}\{o\} \triangleright Owhose[P]are(p)$, and can be given in the query-form CGeomObj whose CPart are CAtrGeom. Similarly the definition from the tool knowledge domain $\Im^{pP}\{o\} \equiv Owith(pR)onP$ CToolObj with CAtrTool on CPart, $\Im^{Wis}\{o\} \triangleright Owhich\_is(a\wedge)$ CToolObj which is CAtrTool $\Im^{WisM}\{o\} \triangleright OwhichIsMadeOf(m\ M\ )$ CToolObj which is CActMake CatAtrTool CatMatTool.

Examples of learning of the basic-forms are presented in Sections 8.2.4.4 and 8.2.4.5.

8.2.4.3 The Text-Task Category – The Explanatory Script

In this section learning of the explanatory script is presented. As it was described in Chapter 7, meaning for the text-task is given by both the explanatory script and the basic-form $\beta(T^{T}) \triangleright (J, B)$. The explanatory script J is used to learn the interpretation of the text-task in terms of the real world objects and the real world situations. The interpretation based on the explanatory script is needed to explain, for example, why some information included in the text is redundant and not useful in solving of the text-task. The explanatory script identifies the categories that are irrelevant to the process of solving the task. For example, the explanatory script of the task "Kate bought 4 apples Each apple cost 2 $ What is total cost" called "buying_type_0" is given as follows: "Kate bought 4 apples Each apple cost 2 $ Total cost is C $ Kate need to pay C $ Kate has enough money to pay Kate spend C $ Kate has 4 apples Kate has less money". Based on this script the explanatory form is learned as follows: (Subject: Kate; Object: apple; Action: buy). The explanatory form is used to identify the irrelevant information and to explain why this information is not needed in the process of finding of the solution.

The explanatory script is learned for each group of the text-tasks that have the same basic-form $T_{i}^{T}(B_{j})$. The first level of the explanatory script includes only the basic explanation. At the first stage of learning only the first level of the explanatory script is learned. The learned explanatory script and the explanatory form are implemented as part of the Explanatory-Expert. The explanatory form is the implementation of the explanatory script that specifies the form of the final result of the explanatory process. The explanatory form for each explanatory level is used to collect the irrelevant information and to explain why this information is not needed in the process of finding of the solution. The Explanatory-Expert is usually invoked in the context of the procedural expert to justify the solution. At this stage of SUS development the explanatory script is learned only for the selected groups of text-tasks.

8.2.4.4 The Text-Task Category – the Mathematical-Text-Task

In this section learning of the mathematical-text-task is presented. The mathematical-text-task $T^T[E(Mat)]$ is the educational text-task $T^T[E]$ that can be found in mathematical texts such as handbooks, supplementary books or examination tests. A mathematical text-task, which has only abstract categories derived from the category of mathematical knowledge, is called a mathematical-abstract-text-task $T^T[E(MatA)]$. A mathematical-text-query, which has only abstract categories derived from the category of mathematical knowledge, is called a mathematical-abstract-text-query $T^G[MatA]$.

A mathematical-text-task that requires finding of the proof of a theorem is called the mathematical-proof-text-task $T^T[E(MatP)]$. A mathematical-text-task that requires explaining of a mathematical problem is called the mathematical-explanatory-text-task $T^T[E(MatE)]$.

The mathematical-text-task usually requires finding of the solution that most often involves two steps: transforming of the text-task into the mathematical formula (e.g. equation) and finding of the solution by solving the equation.

A mathematical text-task can be given as a mathematical-abstract-text-query $T^G[MatA]$ in the form of the symbolic expression without any additional explanation e.g $\left(\frac{2}{5}\right)^2$, $\sqrt{36}+7^2-4$, 20-4=4*(2+?), 2+3; 2+3=; 4*4; 4x5; 3/6; 25+10= or as a mathematical-abstract-text-task $T^T[E(MatA)]$ in the form of the linguistic description "what is half of twenty", or as a mixture of the mathematical expression and linguistic description "what is half of 20". The mathematical-abstract-text-tasks $T^T[E(MatA)]$ do not have categories that have connections with the categories of the real world objects. The simplest mathematical-abstract-text-task is given in the form of description of the operations that need to be performed, eg "Compute 2*7=" or "Solve for x: x + 4 = 7". In solving of the mathematical-abstract-text-task $T^T[E(MatA)]$, at first there is a need to identify the type of task. The simplest mathematical-abstract-text-task $T^T[E(MatA)]$ is given by a mathematical expression and a short description such as "solve equation". The simplest mathematical-text-tasks transformed into a set of categories are used to find the query-form. In order to transform the mathematical-abstract-text-task $T^T[E(MatA)]$ into the query-form the coding category needs to be learned. The coding categories are divided into the mathematical categories and the auxiliary categories. The mathematical categories are derived from the mathematical objects. For the purpose of the research presented in this book, the mathematical category CMath is divided into the numerical category CMathNum, the category of mathematical operators CMathOper or the category of mathematical expressions CMathExpr. The category of mathematical expressions CMathExpr is further divided into the category of algebraic expressions CMathExprAlg, or the category of equations CMathExprEq. The category of mathematical queries is divided into the category CQCompute (compute, find, determine), or the category CQSolve (solve). The category of mathematical objects is divided into the category CMathEquat (equation), the category CMathValue (value), the category CMathBVProb (boundary-value problem), the category CMathParam (parameter), or the category CMathVar (variable).

Learning of the mathematical-abstract-text-task is to learn the query-form, the basic-form and the procedural-form. The explanatory script for the mathematical-abstract-text-task $T^T[E(MatA)]$ explains step by step how the solution of this task was obtained. The explanatory script is learned in the context of the learned mathematical definitions. The query-form is learned based on the mathematical-text-tasks that were selected for learning. These tasks are grouped based on the topic to which they refer. For example, the mathematical-text-task formulated as "Solve for x: x + 4 = 7", "Find solution of equation x + 4 = 7", "What is solution of equation x + 4 = 7", "Which value of x makes equation x + 4 = 7 true", are tasks that refer to the group of tasks marked as "solve equation". For these tasks, the basic-form "solve equation x + a = b" is extracted and learned in the context of the query-form "CQ_SolveLinearEquation". The generalization of the query-form is obtained by analyzing a huge number of mathematical-text-task examples. Examples of learned query-forms: "CQCompute CMathExprAlg", "CQSolve CMathEquatAlg", "CQCompute CMathValue CMathParam CAuxForWhich CMathBVProb CMathExprEqDif CAuxHave CMathSolNTr".

During learning stages, the Math-Task-Expert stores learned knowledge in the form of coding categories, query-forms, basic-forms, procedural-forms and explanatory scripts. The mathematical-abstract-text-task $T^T[E(MatA)]$ consists of the mathematical expressions and linguistic expressions. The mathematical expressions need to be classified into one of the categories of the mathematical expressions such as the category of algebraic expressions, the category of equations (linear, nonlinear, differential, integral), or the category of polynomial transformations. The Math-Expression-Expert implements methods to classify the mathematical expressions into one of the categories of the mathematical expressions. The Math-Expression-Expert invokes the Math-Procedural-Expert that transforms the mathematical expression into the *Mathematica* expression and sends it via MathLink to MathKernel. The description of the Math-Procedural-Expert is presented in Section 8.1.4. The mathematical expression can be also identified utilizing the linguistic description such as "solve equation" that means that mathematical expression is considered to be one of the types of the equations. Most of the text-tasks used for SUS learning were selected from mathematics tests for students [141].

Examples of the text-tasks used for learning of the mathematical-abstract-text-tasks:

$$\text{“}\left(\frac{2}{5}\right)^2,\ \sqrt{36}+7^2-4,\ \text{20-4=4*(2+?), 2+3; 2+3=; 4*4; 4x5”,}$$

“compute 3+6/sqrt(8)”, “determine the values of the parameter L for which the boundary-value problem y’’+L^2y=0 y(0)=0, y(5)=0, has nontrivial solutions”.
“what is next number in this pattern 29 26 23 20 17",
“what is next number in this sequence 29 26 23 20 17",
“what is missing number 20 23 26 32",
“which number is missing from this number sequence 20 23 26 ? 32”,
“which number is between 6857 and 7013 6558 6842 6901 7023";
“which of following numbers has smallest value 0.069 0.2 0.08 0.101",
“which number has smallest value 0.069 0.2 0.08 0.101";
“find number with smallest value 0.069 0.2 0.08 0.101";
“how much greater than 5291 is 5691";
“what is difference between 36323 and 35323".

The text-tasks from the different domains closely related to mathematics, such as statistics, use different categories to describe statistical concepts. The statistical-abstract-text-task $T^T[E(STtA)]$ is the task that includes statistical categories such as mean, median or mode that need to be computed. The statistical-text-task, similarly to the mathematical-text-task, can be formulated by applying only the statistical categories. The learned query-form will also contain only the statistical categories. For example, the text-task “what is mean of these numbers 10 2 4 5 6 0 9 9 " is transformed into the query-form CQ_COMPUTE CStatistic CFor CData”, the basic-form “Compute mean” and the procedural-form “Data [a(1), ,a(n)] compute mean F1: m=(a(1)+…+a(n))/n”. The query-form, the basic-form and the procedural-form are implemented as the part of the Statistical-Expert.

The meaning of mathematical-text-task $T^T[E(MatR)]$ refers to the real word objects or the real world scene. During understanding stage the text-tasks $T^T[E(MatR)]$ are interpreted in terms of the real world situations based on the learned explanatory script. For example, the text-task “Kate bought four apples for $ 2 each what was total cost" refers to the meaning of “buying” category and describes one aspect of the script of “buying”. The script can be given at the different levels of description that show the different aspects of the described scene “buying”. The first level of description of the script “buying” is given as “Man bought n objects Each object cost c $ Total cost is C $ Man need to pay C $ Man has enough money to pay Man spend C $ Man has n objects man has less money”. During learning stage the script is learned along with the query-form. The interpretation based on the script is needed to explain, for example, why some information included in the text is redundant and not useful in solving the text-task. In the case of the mathematical-text-task “Kate bought four apples for $2 each what was total cost" the information such as the name of the subject "Kate" or the name of the objects that are bought "apples", are redundant. The mathematical problem is identified based on the previously learned basic-form and query-form. Learning of the query-form is based on the generalization of the mathematical-text-tasks that have the same meaning. The generalization is concerned with selecting of the coding categories at the proper categorical level. For example, the category "Kate" in the text S1: "Kate bought four apples for $2 each what was total cost" can be selected at the higher categorical level in the form CNameWoman, so all women' names can be properly understood. However, to understand task S1 when the name is "Joe…" there is a need to introduce the category at the higher level "CNameHuman". In the case when text S1 starts with "Teacher…" there is a need to introduce another category CProfesonalName (name of the profession) and the task S1 is written as "(CNameHuman || CatProfesonalName) bought four apples for $2 each what was total cost". Similarly the other categories in the mathematical text-task are coded.

During the understanding stage the coding categories should be decoded into the text categories. For example, CNameHuman is decoded into the proper name "Kate". The categories CNameHuman should be previously learned, that means the term "Joe" is understood as a name if the term "Joe" was learned as a category of CNameHuman (this term is in the categorixicon). In the case when the name "Joe" is not learned the understanding of the text can fail so the learned query-forms needs to include the query-form without CNameHuman category. Testing was performed by applying the text-task that were not used during learning stage. Testing was connected with re-learning the previously learned knowledge.

Learning to solve the mathematical-text-task involves grouping of the mathematical text-tasks depending on the meaning of the texts (the same or similar basic-form). For each group, the query-form, the procedural-form and the interpretational script are learned. During learning, the mathematical text-tasks are grouped as follows: the computational text-task, the text-task that refers to the geometrical model, the "buying" text-task, the "sharing" text-task, the "similar-sharing" text-task, the "time computation" text-task, the "similar-physical" text-task and the "probability" text-task. Each group has the different number of cases (from 20 to 60). In order to learn the query-form, the text-tasks from the groups of the tasks that have similar meaning are selected. The text-task in each group can have the different coding category and the different structure of the text-task. In order to cover the domain of the learning tasks the large sample of the text-tasks is needed for learning. The domain of the text-task is the group of the text-tasks that have the same or similar meaning.

Examples of the "buying" mathematical-text-tasks in which variations of the categories and structures well cover the domain of the learning tasks:

"Kate bought 4 apples for $2 each what was total cost";
"Kate buys 4 new tyres for $12 each what is total cost";
"Kate buys four new tyres for $12 each what was total cost";
"Kate buys 4 new tyres that cost $92 each what was total cost";
"Kate buys four new tyres for $12 each total cost of tyres is";
"Kate buys four new tyres that cost $ ninety two each total cost of tyres is";

"Juice costs $ 3.92 about how much do 6 juices cost",
"Mark buys 4 apples and 5 boxes of juice apples 40 c each juice 65 c per box what is total cost of these items";
"Gemma bought 3 CDs for $ 16 each what was total cost";
"Sam bought 2 apples at 35 c each 3 bananas at 55 c each and 5 mandarins at 25 c each how much money does he spend";
"Sam buys loaf of bread for $ 2.75 and bottle of milk for $ 2.35 how much money does he spend";
"Tina has $ 10.75 she spends $ 6.80 how much money does she have left".
Examples of the "sharing" mathematical-text-task:
"There are 61 books on shelf Peter takes 45 how many books are left";
"6 pizzas are each cut into 6 slices and shared equally between 9 people how many slices does each person get";
"if 20 books are shared equally between Mel Tan Bill and Sara each will get 3 books 4 books 5 books 6 books";
"Jira had two apples she cut each apple into quarters how many quarters did she have in total";
"Ms Watson has bag containing 34 lollies she shares as many of lollies as she can equally between 6 students how many lollies are left over";
"Joe has 40 lollies he gives 19 to Terry and 7 to Kim how many lollies does Joe have left"; "Teacher shares 560 pencils equally among 20 students how many pencils does each student receive";
Examples of the "time computation" text-task:
"Ian left home and spent 0.5 hour walking to swimming pool and then he spent 1.5 hours swimming how long had he been away from home";
"Movie lasting 146 minutes started at 6.30 pm movie finished at";
"Steve sailed his boat continuously from 5 pm 14 July 2003 to 10 am 18 July 2003 for how many hours did he sail";
"Jenny started working at 11.15 am and finished at 12.30 pm same day how long was Jenny working";
"Ben got on bus at 3.46 pm and got off it at 4.12 pm he was on bus for 12 minutes 14 minutes 26 minutes 58 minutes";
"Sue went for run she started at 10.30 am and finished at 11.07 am for how many minutes was Sue running".

8.2.4.5 The Text-Task Category – The Physical-Text-Task

In this section learning of the physical-text-task is described. A text-task from physics called a physical-text-task $T^T[E(Phs)]$ is the task that refers to the physical object or to physical phenomena. The real world phenomenon is represented by the physical model that is built based on abstraction process. The physical model is given usually in the form of the mathematical equation where variables refer to specific categories of the physical knowledge. In this book we will refer to the very small part of the physical knowledge that deals with description of the movement of the object with constant velocity. The physical model is described using the physical or mathematical categories such as velocity, time or distance.

The physical-text-tasks were learned based on the problems presented in books that are used for help in mastering skills of students in solving problems [138], [139], [141]. Many books from physics offer one model problem and many problems following the model to practice the skills in solving a given problem. Similarly SUS learns the physical-text-task by selecting the tasks which have the same or very similar meaning represented by the basic-form. At first, the simple text-tasks are learned and next the text-task that are more difficult in the sense of difficulty of understanding of the categories $\delta[C]$ and the difficulty of understanding of the query-form $\delta[\theta]$. The text-task difficulty level is described in Section 7.3.2.1.

At first, physical-text-task selected for learning physical-text-task are grouped into groups that have the same basic meaning and each physical-text-task $T^T[E(Phs)]$ is transformed into the basic-form $\beta_j(T^T[E(Phs)]) = B_j$. Next, the physical-text-tasks are grouped into groups that have the same basic-form $T_i^T \in T^T[E(Phs)](B_j) if \beta_j(T^T[E(Phs)]) = B_j$. These groups are arranged according to the text-task difficulty level. Each group of physical-text-tasks $T^T[E(Phs)](B_j)$ is further grouped into groups that have the same query-form $T^T[E(Phs)] \in T^T[E(Phs)](B_j(\theta_k)) if \Theta_k(T^T[E(Phs)](\beta_j)) = \theta_k$ and the query-form $\Theta_k(T_i^T)$ is learned for each group of physical-text-task $T^T[E(Phs)](B_j(\theta_k))$. In this Chapter the text-task and learning of the physical text-tasks that refer to the topic "movement of the objects", is described.

In order to learn the query-form, the basic-form and the procedural-form of the physical-text-task the examples of the tasks were selected from [138], [139]. Examples of the selected physical-text-task used to explain learning methods are as follows:

Z1 "car travel 10 km in 30 min what was his speed";
Z2 "student travels 10.0 km in 30.0 min what was her speed",
Z3 "student driving car travels 10.0 km in 30.0 min what was average speed";
Z4 "car moves along a road at average speed of 5 m/s how far will it travel in 15 minutes";
Z5. toy train moves along winding track at average speed of 0.25 m/s how far will it travel in 4 minutes";
Z6 "robot that left its closet and traveled 1200 m had average speed of 20.0 m/s how long did trip take";
Z7 "rolling across machine shop at constant speed of 4.25 m/s robot covers distance of 17.0 m how long did that journey take";
Z8: "how long does it take a train to travel the distance of 50 km at speed 42 km/h";
Z9: "how long will it take a John to run the distance of 2000 m at velocity 7 km/h".

In order to transform the physical-text-task into the query-form, at first the categories of the specific domain need to be learned. The categories are learned in the context of the explanatory script that depicts the selected aspects of the "moving object" topic. The moving object script depends on the category of moving object and is different for the moving vehicle, moving animal or moving astronomical object. For example, the moving vehicle (car) script includes the following activity: "car is parking, car is in garage, driver is going into car, driver start the car, car is moving from garage (parking), car is on the road, car is moving with constant speed, car is moving into garage, driver stop the car, car rest, driver is going out". In the case of the task Z1 "car travel 10 km in 30 min what was his speed" the script supplies knowledge about the car "traveling". The information given in this example is not "complete" and SUS needs to assume that someone is driving the car (car usually can not travel without driver). Our understanding easily "fills the gap" so we do not notice that this information is "incomplete".

In order to transform the physical-text-tasks Z1-Z9 into the query-form, at first the coding categories of the specific domain need to be learned. The coding categories are divided into the mathematical categories, the physical categories and auxiliary categories. It is assumed that the coding categories are learned from the sample of physical-text-tasks used for learning.

The numerical category CNum is the category derived from the mathematical objects. The category of speed units [CUv], the category of time units [CUv], and the category of distance units [CUs] is divided into the category of word units and the category of symbol units, and includes the category of speed units [CUWv] (kilometre per hour, meter per second), the category of time units [CUWt] (minute, hour, second) and the category of distance units [CUWs] (kilometre, meter).

The category of symbolic units includes the category of speed units [CUSv] (km/h, m/s), the category of time units [CSUt] (m, h, s) and the category of distance units [CUSs] (km, m).

The physical categories CPhisObj include:
the category of speed CPhisObjSpd (speed, velocity),
the category of average speed CPhisObjAvrSpd (average speed, average velocity),
the category of distance CPhisObjDist (distance), and
the category of time CPhisObjTime (time).
The category of moving objects CObjMov is divided into:
the category of man CObjMovMen (toy train, man, women, child, children, kids, boy, girl, teenager, schoolgirl, daughter, son, student, father, mother, worker, skier, runner, cyclist, driver, rider, horse rider, John, Sue, student driving car),
the category of moving animals CObjMovAnim (dog, horse), and
the category of vehicles CObjMovVieh.
The category of vehicles is further divided into:
the category of earth vehicles CObjMovViehErt (bicycle, motorcycle, car, bus, truck, tractor, tank, trolley-bus, diesel train, electric train, high-speed train, tram),
the category of air vehicles CObjMovAir (balloon, airplane, plane, airship, glider, helicopter, rocket), and

the category of water vehicles CObjMovWat (raft, amphibia, boat, canoe, kayak, ship, steamship, longship, ocean liner, junk, jetfoil, galley, galleon, clipper ship, motor-boat, yacht, submarine, bark, sailboat, caravel).

These categories can be coded at the lower level, for example, the object from the CObjMovViehErt category can be coded as the car category CObjMovViehErtCar or by using the car name as Ford or Toyota Altice.

Living object is moving using legs or crawls so the category of moving part of the object describes which part of the moving object contributes to the moving object: man (legs), animals (legs), bird (legs, wings), car (wheel, engine).

The category of moving actions CActMove is divided into the category of moving on the earth CActMoveEarth (moving, travelling, walking, running, crawling, jogging, jumping, riding, driving, sliding, rolling, covering), the category of moving on the water CActMoveWat (swimming, boating, sailing, steaming, floating) and the category of moving in the air (flying, rising). The category of moving actions is connected with the category of moving objects and is given as the category of moving object-action. For example, the category of walking objects (man, animal, robot), the category of flying objects (bird, kite, airplane).

The text-task usually contains the redundant information that is not needed in order to find the solution. This information is coded as the background-coding-categories and is used during explanatory process. For example, the category of types of moving tracks includes CTrMove (winding track, oval track, road, path, freeway, pavement, air, river, sea, ocean, lake). The category of the type of moving tracks is associated with the category of moving places that includes CMovePl (town, center of town, lawn, forest, market, machine shop, shop, office, corridor, parking, parking lot, steel water on lake, flowing water of a stream, upstream, downstream, closet, hallway in a space terminal,). The category of directions of moving includes CDirMove (along, across, eastward, southward, east). The category of changes of movement includes CChangMove (turn due south, end). The category of changes of movement is associated with the category of types of beginning state CBeginMove (start from rest, come from home, come from garage, left its closet, left building (home, office, shop), left the garage parking, launch a rocket that rises into the air, starting from the centre of town, initially moving). The category of types of measurement includes CMeasure (according to its computer, trip meter to record distance, car odometer, watch, stoper). The category of circular movements includes CTypMoveCir (makes one lap, around N-m track, around oval track). The category of queries includes CQCalcul ={find, calculate, what is, express, compute}, CQHowFar (how far), CQHowLong (how long). The auxiliary categories can be CAuxIn (in), CAuxAt (at), CAuxHave (have), CAuxTake (take).

The query-forms for the text-tasks Z1-Z6 are composed from the learned coding categories and have the following query-forms:

Z1, Z2: "CObjMov CActMove CNum [CUv] CAuxIn CNum [CUt] CQCalcul CatPhisObjSpd",
Z3: "CObjMov CActMove CNum [CUv] CAuxIn CNum [CUt] CQCALCULATE CatPhisObjAvrSpd",

Z4: "CObjMov CActMove CDirMove CTrMove CAuxAt CatPhisObjAvrSpd CNum [CUv] CQHowFar CAuxIt CActMove CAuxIn CNum [CUt]",
Z5: "CObjMov CActMove CDirMove CTrMove CAuxAt CatPhisObjAvrSpd CNum [CUv] CQHowFar CAuxIt CActMove CAuxIn CNum [CUt]",
Z6: "CObjMov CAuxThat CBeginMove CAuxAnd CActMove CNum [CUs] CAuxHave CatPhisObjAvrSpd CNum [CUv] CQHowLong CatPhisObjTrip CAuxTake",
Z7: "CActMove CDirMove CMovePl CAuxAt CatPhisObjConsSpd CNum [CUv] CObjMov CActMove CatPhisObjDist CNum [CUs] CQHowLong CatPhisObjTrip CAuxTake",
Z8, Z9: "CQHowLong CAuxTake CObjMov CActMove CatPhisObjDist CNum [CUs] CAuxAt CNum [CUv] CatPhisObjSpd".

The basic-form represents the meaning of the text-task and the different query-forms such as the query-form for the tasks Z1, Z2, Z3 can refer to the same basic-form. The text-tasks Z1, Z2, Z3 are transformed into the basic-form: "Object O move distance s [s] in time t [t]. Compute velocity v [v]".

The text-tasks Z4 and Z5 are transformed into the basic-form: "Object O move with velocity v [v] in time t [t]. Compute distance s [s]".

The text-tasks Z6, Z7, Z8 and Z9 are transformed into the basic-form: "Object O moves with velocity v [v] distance s [s]. Compute time t [t]".

The tasks Z1, Z2, Z3 refer to the procedural type of data s [s], t [t] find v F: v=s/t [v], the tasks Z4, Z5 refer to the procedural type of data v [v], t [t] find s F: s=v*t [s], whereas tasks Z6, Z7, Z8 and Z9 refer to the procedural type of data s [s], v [v] find t F: t=s/v [t]. The procedural type of data that makes it possible to solve the text-task is obtained from the basic-form. The basic-form includes only the physical categories that are associated with data and indicate the physical model (mathematical formulae) used to compute the solution. The computation of the distance, velocity or time after transforming the text-task into the basic-form is not the difficult task, however to translate the text-task into the query-form can be in some cases a quite difficult problem to solve. The text-task includes categories of real world objects so the text-task can have very different forms. Understanding these text-tasks requires excessive learning in order to cover all possible cases of the text-task. Similarly, like in solving mathematical problems described in previous part, understanding of the deep meaning of the text-task refers to the explanatory script that depicts the scene of a moving object. Understanding of the deep meaning of the text-task is not needed to solve the text-task, however it can help to explain very precisely the importance of each category in finding the solution.

Learning how to solve the physical text-task is to learn of the query-form, the basic-form, the procedural-form and the explanatory script. Here is the example of the query-form, the basic-form, the procedural-form and the explanatory script for the following text-task: "auto travels at rate of 25 km/h for 4.0 minutes and finally at 20 km/h for 2.0 minutes find total distance covered in km".\

The query-form: CObjMov CActMove CAuxAt CNum [CUv] CAuxFor CNum [CUt] CAuxAnd CAuxEnd CNum [CUv] CAuxFor CNum [CUt] CQFind CDistance CAuxIn [CUs].

The basic-form: Object move at speed v1 in time t1 and next at speed v1 in time t1 Find total distance in sU.

The procedural-form: Data v1 [vU], t1 [tU] v2 [vU], t2 [tU]; Find: s. Defining equation: v=s/t, Convert units: v[vU]=v[m/s], t[tU]=t[s]. Total distance equals the sum of the distances covered during period t1 and t2, so s=s1+s2=v1*t1+v2*t2.

Learning how to solve the physical-text-task requires learning variations of the text-tasks from the different domains of physics. The physical-text-tasks refer to the well defined model so the difference in the learning text-task from the different domains of physical knowledge depends on the different physical categories and models. In this book two groups of the text-tasks are presented: the motion with constant velocity and the uniformly accelerated motion. The physical model is defined using the physical categories such as acceleration, velocity, time or distance. The selected text-tasks have huge structural variations and huge variations of the different categories. Here are examples of the text-tasks used for learning how to solve the physical-text-task "the motion with constant velocity":

TT1: "John travels 120 km in 8 hours average speed of his is";
TT2: "cyclist travels 120 km in 8 hours average speed of cyclist is";
TT3: "student driving car travels 10.0 km in 30.0 min what was his speed";
TT4: "snail travels 45 cm in 20 min it average velocity is";
TT5: "car travel twenty kilometers in two hours what is its average velocity";
TT6: "student driving car travels 10.0 km in 30.0 min what was her average speed
TT7: during race on oval track car travels at average speed of 200 km/h determine its average velocity at end of its third lap";
TT8: "runner makes one lap around 200 m track in time of 25 s what was runner average speed and average velocity";
TT9: "John makes one lap around 200 m track in time of 25 s what was his average velocity";
TT10: "runner travels 1.5 laps around circular track in time of 50 s diameter of track is 40 m what was runner average speed";
TT11: "runner travels 1.5 laps around circular track in time of 50 s circumference of track is 126 m find average speed of runner";
TT12: "runner travels two laps around circular track in time of 50 s diameter of track is 40 m and its circumference is 126 m find average speed of runner";
TT13: "auto travels at 25 km/h for 4.0 minutes and finally at 20 km/h for 2.0 minutes find average speed for complete trip in m/s";
TT14: "car's odometer reads 22687 km at start of trip and 22791 km at end trip took 4.0 hours what was car's average speed in m/s";
TT15: "kids launch rocket that rises into air along 380 m arc in 40 s determine its speed";
TT16: "three kids in parking lot launch rocket that rises into air along 380 m long arc in 40 s determine its average speed";
TT17: "auto travels at 25 km/h for 4.0 minutes and finally at 20 km/h for 2.0 minutes find average speed for complete trip in m/s
TT18: car travels at 100 km/h for 2 h at 60 km/h for next 2 h and finally at 80 km/h for 1.0 h what is car's average velocity for entire journey".

The tasks that vary only in one word can have usually the same meaning. Words that refer to the category of moving object such as John, snail, cyclist, and student driving car can be easily "filtered" by learned query-form. Also other categories that give the redundant information can be easily recognized by learned query-form. In learning of the query-form the large sample of learned examples makes it possible to obtain the proper generalization. However, the explanatory script that is used to explain the redundancy of the words needs to learn the more specific query-form that makes it possible to identify redundant categories and supply the proper explanation.

8.2.4.6 The Text-Task Category – The Geographical-Text-Task

In this section learning of the geographical-text-task is described. The geographical-text-task $T^T[E(Geo)]$ is the educational text-task $T^T[E]$ that can be found in geographical texts such as handbooks, supplementary books or examination tests. The geographical-text-task refers to the domain of geographical knowledge. The geographical text-queries $T^G\{Geo\}$ are geographical queries that refer to geographical domain and that can be usually found in the crossword puzzles. These queries can contain the short description e.g "Norwegian south pole explorer".

Learning of the geographical-text-queries $T^G\{Geo\}$ or geographical-text-task $T^T[E(Geo)]$ is connected with learning of the geographical knowledge. In contrast to the mathematical-text-task or the physical-text-task, the most of the geographical-text-tasks belong to the category of search problems. As it was described, the basic-form of the category of search problems refers to the search query that is used to find required geographical facts. The query-form and the basic-form are learned based on the selected geographical tasks. Learning of the query-form requires learning the coding categories. The geographical coding categories are part of the geographical knowledge. As it was described, the basic geographical categories are categories of geographical objects related to the categories of geographical domains. The category of geographical objects G_i includes the category of mountain, the category of peak, the category of fiord, the category of river, the category of lake, the category of sea, the category of ocean or the category of island: CGeOBJ ={Criver, Clake, Csea, Cocean, Cisland, Cmountain, Cpeak, Cfiord}, whereas the category of geographical domains D_j includes the category of continent, the category of country or the category of province CGeDOM ={Cworld, Ccontinent, Ceurope, CPoland, CVictoria). The coding categories of the geographical object are represented as $Co = [Criver,\ Clake,\ Cmountain,\ Cisland,...]$ and the coding categories of the geographical domains are represented as the hierarchical structure $X^0 = CEarth \quad X^1 = Ccontinent \quad X^2 = Ccountry \quad X^3 = Cdivision0$, where $X^1 = Ccontinent = [CEuropa,\ CAsia,\ CAfrica,\ CSouth\ America,,...]$,

$$X^2 = Ccountry = [CEngland,\ CAustralia,\ CPoland,\ CUSA\ CPeru,,...],$$
$$X^3 = Cdivision0 = [CVictoria,\ CPensylwania,\ CMalopolska,,...].$$

The simple geographical-text-queries are represented by the category of geographical objects G_i and the category of domain objects D_j and are denoted as $T^G\{Geo\}[G_i]\{D_j\}$. The geographical text-queries that refer to the specific geographical objects located in the specific geographical domains are learned as the query-form $\Phi(T^G\{Geo\}[G_i]\{D_j\})$, the basic-form $B(T^G\{Geo\}[G_i]\{D_j\})$ and the procedural-form $P(T^G\{Geo\}[G_i]\{D_j\})$. The text-queries $T^G\{Geo\}[G_i]\{D_j\}$ are represented in the query-forms as follows: $\Phi(T^G\{Geo\}[G_i]\{D_j\})$ GeOBJinDOMAIN e.g. RIVERinCOUNTRY or MUNTAINinCONTINENT. For example, the text-queries "river in Poland", "mountain in Hungary", "lake in Armenia", "island in France" are transformed into the query-form GeOBJinDOMAIN and into the basic-form $o\_in\_X^i_j$ (object o in the container X^i_j). As it was described in Section 8.1.4, the procedural-form $P(T^G\{Geo\}[G_i]\{D_j\})$ specifies the learning procedure that stores geographical facts and the searching procedure that makes it possible to search for these facts during understanding process. The learning procedure and searching procedure are strictly connected and specify the mechanism of storing and accessing knowledge. The learning and searching procedure can employ the different storing and searching methods that make it possible to optimize learning and understanding process. For example, learning the basic-form given by the query-form GeOBJinDOMAIN requires learning all geographical objects in all general and specific geographical domains (containers). One way of learning is to select the object o_k such as the river and to learn specific geographical facts on each categorical level represented by container X^i_j. For example, by selecting the categorical level X^1_j (continent) all rivers on each continent are learned $L(X^1_j) \triangleright [r^1_1,...,r^1_k],\ where\ j = 1,...,M$, where M is a number of continents. By selecting next categorical level X^2_j (country) $L(X^2_j) \triangleright [r^2_1,...,r^2_k],\ where\ j = 1,...,H$ (H is a number of countries), knowledge of the next categorical level is learned. As the result, the learned geographical facts are stored in the form that is easily accessible by the query e.g. river in Europe or river in Poland. Learning procedures provide a very simple mechanism of the checking of the correctness of the learned facts.

During learning, the simple text-query $T^G\{Geo\}[G_i]\{D_j\}$ is transformed into the text-task $T^T\{Geo\}[G_i]\{D_j\}$ by adding the query-part. The text-part can be learned independently so the learning the text-task $T^T\{Geo\}[G_i]\{D_j\}$ consists of learning of the text-part (text-queries $T^G\{Geo\}[G_i]\{D_j\}$) and the query-part. Knowledge for the simple text-queries $T^G\{Geo\}[G_i]\{D_j\}$ is stored in the text format as GeoObjDomcategorexicon file. During learning of the geographical knowledge the Geographical-Expert invokes the Geographical-Query-Form-Expert, the Geographical-Basic-Form-Expert and the Geographical-Procedural-Form-Expert to learn the query-form, the basic-form and the procedural-form. During solving of the geographical problem the Geographical-Query-Form-Expert invokes Geographical-Basic-Form-Expert and the Geographical-Procedural-Form-Expert. The Geographical-Query-Form-Expert implements the query-form as clauses that invoke the Geographical-Basic-Form-Expert. The Geographical-Procedural-Form-Expert invokes searching algorithms to search GeoObjDomcategorexicon.

The text-queries such as "country in Europe", "province in Belgium", "county in Ireland", "capitol of Belgium", "town in Poland" are represented by the domain object D_j and are given as $T^G\{Geo\}[D_i]\{D_j\}$, where both the object and the container are domain objects D. The basic-form $B(T^G\{Geo\}[D_i]\{D_j\})$ is given as $x\_in\_X^i_j, where\ x \in X^k_j,\ k > i$, where x is a category at the level k that is lower that the level i.

The type text-query $T^G\{Geo\}[typ](C_i)$, such as "type of stream" is learned by transforming it into the query-form $\Phi(T^G_Q[typ])C_i))$ "typeObj". Geographical knowledge for the query-form of the type text-query $T^G\{Geo\}[typ](C_i)$ is learned for each category C_i and is stored in the text format as Geotypecategorexicon file. Similarly, the part text-query $T^G\{Geo\}[part](C_i)$ is learned for each category C_i and is stored in the text format as Geopartcategorexicon file. The geographical attribute text-queries $T^G\{Geo\}[C_i](p\ )$ are learned by transforming it into the query-form $\Phi(T^G[C_i](p_1,...,p_N))$. Similarly, attribute text-query $T^G\{Geo\}[C_i](p\ )$ is learned for each attribute $p_1,...,p_N$ and is stored in the text format as GeoAttributecategorexicon file.

The geographical text-queries with qualitative attributes $T^G\{Geo\}[[G_i](p\ )]\{D_j\}$ such as greater, small or higher are compound names of the category derived from the general category, usually combination of the name of the marker and name of category of the geographical object e.g small river in Europe. The marker such as "highest" refers to the name of the one geographical object and in this case the text-query is the synonym for the specific name of the derived category. For example, the highest peak is the synonym for Mount Everest. These fact that are given by one case such as "largest fiord in Norway", "biggest river in Europe", "capital of Romania" can be learned independently or as a list of categories of the higher level, for example, "Sofia capital of Bulgaria" or CountryCAPITAL.

The text-query with the specific category derived from the river category based on the attribute "size" is the "small river in Europe" that has query-form CSmallRiverInEUROPE. The small river categories can be stored as the list of river names ordered from the biggest to the smaller one or can be dynamically created based on the value of the attribute PrSize. The text-queries $T^G\{Geo\}[G_i\ ]\{D_j\}([C_i\ ](val(p\ ) = c^i)$ are represented in the query-form $\Phi(T^G\{Geo\}[G_i\ ]\{D_j\}([C_i\ ](val(p\ ) = c^i))$ e.g. BigRIVERinCOUNTRY or HighMOUNTAINinCONTINENT. The general query-form AtrPGeOBJinDOMAIN has its basic-form $AtrP(o)\_in\_X_j^i$ (attribute P object of object o, $AtrP(o)$ in the container X_j^i). The procedural-form specifies the learning and the searching procedure. For example, learning of the basic-form given by the query-form AtrPGeOBJinDOMAIN require to learn all derived geographical objects in all general and specific geographical domains (containers). Learning is similar, like in the case of geographical objects in the container, however in this case there is a need to learn also the values of attribute $val(p\ )$. Another way is to learn the selected object o_k such as the river in selected order at each categorical level represented by container X_j^i. By selecting categorical level X_j^1 (continent) all rivers on each continent are learned $L^O(X_j^1) \triangleright [r_1^1,...,r_k^1],\ where\ j = 1,...,M$ where M is a number of continents, and sorted in ascending order. As the result the learned geographical facts are stored in the form that is easily accessible by Geographical-Procedural-Form-Expert.

The text-queries such as "country with high population" are transformed into the basic-form $x\_with\_high\_a,\ where\ x \in X_j^k$ and $high\_a$ denotes the value of attribute a. The value of the attribute a is usually given in the interval $high\_a \equiv [a:\ val(a) \in (a1, \max)]$, where max is a max value of the attribute a. The text-queries such as "country with highest population" are transformed into the basic-form $x\_with\_\max\_o,\ where\ x \in X_j^k,\ k > i$ and max is a highest value of the attribute a. The text-queries such as "country in Asia with biggest area", "country in Asia with highest population" or "capitol of Europe with highest population" are transformed into basic-form $x\_in\_X_j^i\_with\_\max\_p,\ where\ x \in X_j^k,\ k > i$. Similarly text-queries such as "river of Sweden that drains into Baltic Sea" are transformed into basic-form $x\_of\_X_j^i\_that\_v\_S\ ,\ where\ x \in X_j^k,\ k > i,$ and $v = [drain\_into,\ flowing\_into,\ emptying\_into]$ is a coding category of river end and $S = \{Baltic,\ Northern,...\}$ is the coding category of ocean or sea.

Similarly, the text-query "length of Vistula river" is transformed into the basic-form $p\_N(o)$ and the text-query "length of longest river in Poland" is transformed into $\max p\_o\_in\_X$ and "population of capitol of Poland" into $o\_of\_o\_of\_X$. For each basic-form the procedural-form that contains the searching procedure such as $Find\_x\_with\_high\_a()$ is learned.

The learned geographical knowledge is stored in the Excel format during learning process and is converted into the text format during understanding process. The Excel format is used for easy access to the data during learning process. An example of learned geographical knowledge converted into the text format is shown in Table:

world	population	world	density
Germany	82689210	Taiwan (Republic of China)	636
United Kingdom	60776238	South Korea	493
Ukraine	46480700	Lebanon	344

The most important part in learning process (knowledge implementation) is to design a relatively simple method for checking the correctness of learned knowledge and to implement fast easy access during understanding process. The Geographical-Task-Expert uses the text format that is easy to process during searching of a particularly type of data. The Excel format is used mostly during learning whereas the text format is used during understanding process. During understanding process the Geographical-Task-Expert invokes the Geographical-Procedural-Form-Expert to solve the problem. The Geographical-Procedural-Form-Expert implements searching algorithms to search Geocategorexicon.

As it was shown, the text-queries can be easily converted into the text-task. For example, the text-queries: "tributaries of Vistula river" can be converted into the text-task "give name of right tributaries of Vistula river", "what are right tributaries of Vistula river" or "list all right tributaries of Vistula rivers", refer to the basic-form "right tributaries of Vistula river" and are given by the query-form "CQ_WhatIs CPrLOCATE CPartRIVER Cof CNameRIVER CRIVER" or in the general form "CQ_WhatIs CPrLOCATE CGePartOBJ Cof CGeName CGeOBJ". The coding category CQ_WhatIs={give name of || what IS || list all, …} includes all possible question parts that can be combined with the basic-form to give the query-form. For example, the text-query "Accra is capital of which African country" has a query-form "CNameCapitol CIs CCAPITOL Cof Cwhich CPrCONTINENT CCOUNTRY" or "CGiveName Cof Capitol CCAPITOL Cof Cwhich CPrCONTINENT CCOUNTRY".

The text-queries such as "river in Poland" or "mountain in Hungary" can be expressed by using predicate categories PrCountry={Polish, Hungarian, …}, PrContinent={European, American, …} and transformed into the text-task query-form "CQ_WhatIs CPredCountry CGeoOBJ" or "CQ_WhatIs CGeoOBJ Cin CGeDOM".

Learning to solve the geographical-text-task requires learning the geographical facts, the query-form and the procedural-form (searching procedure).

The Geographical-Procedural-Form-Expert implements the searching procedure that has access to the learned geographical facts stored in the in text file format. Geographical tasks are learned by SUS by learning variation of the text-queries selected from the different crossword puzzles or geographical tests. The text-queries were divided into two groups. The first group was based on the definitions of the geographical objects such as a river, lake, sea, ocean, island, mountain, peak, fiord, soil, erosion. For example, "type of lakes", "part of river". The second group contained the geographical-text-tasks and the geographical-text-queries represented by the geographical object and the domain object. During the learning stage the query-form was obtained during generalization process from the selected text-queries.

During learning, at first, for the different geographical object categories such as CRiverOBJ, CMountainOBJ, CLakeOBJ, CSeeOBJ, CIslandOBJ the containers such as CContinentDOM, CCountry, CProvinceDOM were selected. The object categories and container categories are learned as the coding categories used to learn the query-form. At the first stage of learning the selected combinations of these categories as the query-form were learned. For the query-form such as CatRiverOBJ in CatContinentDOM, the rivers for each continent were learned by storing names of the rivers in the text file. The procedural-forms that make it possible to store these geographical facts and searching procedures that make it possible to search for these facts during understanding process are learned independently.

During learning process Geographical-Task-Expert is invoked to learn the query-form $\Phi(T^G\{Geo\}[[G_i](p)]\{D_j\})$, the basic-form $B(T^G\{Geo\}[[G_i](p)]\{D_j\})$ and the procedural-form $P(T^G\{Geo\}[[G_i](p)]\{D_j\})$. During solving the geographical problem with attribute text-queries the Geographical-Task-Expert invokes the Geographical-Procedural-Form-Expert to solve the problem. The Geographical-Procedural-Form-Expert implements searching algorithms to search Geoattributecategorexicon. The qualitative attributes such as size has its value given as the set of categories $val(p) = \{c^1,...,c^N\}$.

During solving the geographical problem the Geographical-Query-Form-Expert invokes Geographical-Basic-Form-Expert and the Geographical-Procedural-Form-Expert. The Geographical-Query-Form-Expert implements the query-form as clauses that invoke the Geographical-Basic-Form-Expert. During solving the geographical problem the Geographical-Procedural-Form-Expert that invokes searching algorithms to search geotypecategorexicon is invoked. During understanding process the Geographical-Task-Expert invokes the Geographical-Query-Form-Expert to analyse the query and the Geographical-Procedural-Form-Expert to invoke an appropriate search method such as $Find\_o\_in\_(O_h, X_j^i)$, where geographical object O_h and container X_j^i are specified by the Geographical-Basic-Form-Expert. For example, during understanding text-query "river in Poland" the solution is given as a set of categories $[r_1^i,...,r_k^i]$ represented by rivers names. In some cases the solution can be too specific, a numbers of rivers can be too high so the Geographical-Explanatory-Expert will optimize the solution in the context of the contextual information.

8.2.4.7 The Text-Task Category – The Tools-Text-Task

In this Section learning of the tools-text-task is presented. The tools-text-task $T^T[Tol]$ is the text-task T^T that can be found in books that contain knowledge about tools. Similarly, the tools-text-query $T^G\{Tol\}$ is the text-query that refers to the knowledge about tools. The crossword puzzles queries about variety of tools provide many examples of the tools-text-queries. These queries can contain a short description e.g. "nail with small flat head" or more complex description such as "slender metal shaft that is pointed at one end and flattened at other end". In contrast to the educational text-task that refers to the scientific knowledge, the tools-text-task or tools-text-query most often refers to the common sense knowledge. The tools-text-task or the tools-text-query are usually learned from tools definitions expressed in the form of the short description.

In order to learn knowledge from a tools domain there is a need to learn the definition that describe the different aspects of the defined object. The definition that refers to the selected object not only describe the different aspects of the defined object but also can be differently formulated when refer to the same aspect of the object. The complex definitions can be transformed into the simple form by omitting some categories or some aspects of the definition, for example "nail that is similar to common nail but has slimmer shank and is used on lighter pieces of wood and on boxes" can be transformed into simplest form: "nail that is similar to common nail but has slimmer shank", "nail that is similar to common nail and is used on lighter pieces of wood and on boxes", "nail that is similar to common nail but has slimmer shank and is used on lighter pieces of wood", "nail that has slimmer shank and is used on lighter pieces of wood and on boxes", or "nail that is used on lighter pieces of wood and on boxes".

Learning of the knowledge of the tools-text-task is to learn the coding categories, the query-form, the basic-form and the procedural-form. The first step in learning of the tool-definition-text-query is to learn the coding categories and the basic-form and the query-form. The query-form is the generalization of the text-task expressed in the form of the clauses consisting of the coding categories. The important part of learning of the query-form is generalization based on the selected examples of the text-task. Learning of the procedural-form requires learning facts that are "answers" to the query. For example, for the tool-text-query: "tool used for blowing chisel" the learned fact is "hammer".

The tools-definitions-text-queries from the tools domain are learned using examples of the different forms of tools queries such as the consist-text-query, the made-text-query, the used-material-text-query, the used-result-text-query, the used-as-text-query, or the used-for-text-query. These tools-definitions-text-queries are described in previous chapters. The tools-text-queries that belong to the tool domain are learned based on the selected queries of the crossword puzzles. The tools-text-task are obtained from the tools-text-queries by adding the query-part.

The tools-text-tasks that are learned provide tools knowledge needed during understanding process. Understanding of the texts of the tools domain is tested in the context of solving crossword puzzle queries. Learned knowledge is tested by answering queries similar to previously learned tools-text-task by alternating words and using different definitions for the same object category. Here are examples of tool-text-task that are tested by alternating the words "Q slender metal shaft that is pointed at one end and flattened at other end"; "Q slender shaft that is pointed at one end and flattened at other end"; Q metal object that is pointed at one end and flattened at other end"; "Q slender object pointed at one end and flattened at other end"; "Q object that is pointed at one end and flattened at other end"; "Q slender metal object pointed at one end". The symbol Q denotes the query-part that can be added to the query-form to obtain the text-task.

Learned tasks are grouped into groups that make it possible to obtain generalization of these tasks in the context of learning of the query-form. In contrary to learning of the text-task (text-queries) from scientific domains such as mathematic or physic that are related to well defined knowledge learning of the text-task from the tools domains need to cover broader spectrum of the text-task. The variation of categories and variation of the structure of the task require learning the big sample of tasks to obtain proper generalization of the query-form. Example of tasks used in learning and testing tools-text-task that are given in this section shows grouping needed during learning of the query-form. The important part of learning of the query-form is to learn coding categories. The coding categories should be selected at the categorical level that does not lead to errors caused by mixing text-task that have similar meaning but refers to the different categories of objects.

Examples of the tools-text-tasks used during learning of the query-form:

"what is name of nail that has small flat head"; "what is name of nail that has round crossection and small flat head";

"what is name of nail with head wider than shank"; what is name of nail that has head scarcely wider than shank"; "what is name of nail that has head scarcely wider than shank";

"what is name of nail that has galvanized finish"; "what is name of nail that has coppered finish"; "what is name of nail that often has coppered finish"; "what is name of nail that has zinc or galvanized finish"; "which nail has zinc or galvanized finish";

"what is name of nail that has round crossection"; "what is name of nail that is round in crossection"; what is name of nail that has flat shank"; "what is name of nail that has diamond-shaped head"; "what is name of nail that has diamond-shaped head that is driven in flush with board surface";

"what is name of nail that has jagged shank to grip framing to which board is being fixed"; "what is name of nail that has jagged shank"; "what is name of nail that has cone shaped head"; "what is name of nail that has cone shaped head and jagged shank"; "what is name of nail that has cone shaped head and jagged shank to grip framing"; "what is name of nail that has cone shaped head like countersunk screw"; "what is name of nail that has cone shaped head like countersunk screw and jagged shank"; "what is name of nail that has cone shaped head like countersunk screw and jagged shank to grip framing"; "which nail has cone shaped head like countersunk screw and jagged shank to grip framing"; "which nail has cone shaped head like countersunk screw and jagged shank to grip framing to which board is being fixed".

8.2.5 Implementation of the New Knowledge – The Dictionary-Text

In this section learning of the dictionary-text is described. As it was described in Chapter 7, the dictionary-text T^D is the short text found in dictionaries or in encyclopedias. Learning of the dictionary-text T^D requires learning of the query-form, the basic-form, the procedural-form and the interpretational script. The dictionary-text given in the form of the encyclopedic or dictionary definition is called the definition-dictionary-text $T^D[Def]$ and has no more than one sentence. The different dictionary-texts contain knowledge from the different domains of the scientific discipline such as physic or chemistry. The definition-dictionary-text $T^D[Def]$ is learned in the context of the physical-dictionary-text $T^D[Phy]$, the chemical-dictionary-text $T^D[Chm]$, the biological-dictionary-text $T^D[Bio]$, or the mineralogical-dictionary-text $T^D[Min]$. The biographical-dictionary-text $T^D[Big]$ is learned in the context of the mineralogical-dictionary-text $T^D[Min]$ as a part of the interpretational script connected with the name of the discoverer Λ_i^N. The literature-dictionary-text $T^D[Lit]$ is much more difficult to understand and to learn than other dictionary-texts.

The dictionary-text T^D is learned in the context of the interpretational script $S(T^D)$. The interpretational script $S(T^D)$ is given by the model of the phenomena $M(S_i)$ to which the script refers. The model $M(S_i)$ is build based on all accessible knowledge and is related to the level of understanding. At the basic level of understanding the model $M(S_i)$ includes the information that is needed to understand the learned dictionary-texts. That means, that only those dictionary-texts will be understood for which the query-form, the basic-form and the interpretational script were learned.

The dictionary-texts T_i^D used for learning are prearranged into groups that have the same basic meaning and for each group the basic-form $\beta_j(T_i^D) = B_j$ is learned. The basic-form B_j needs to be consistent with the model $M(S_i)$. The consistency means that basic-form B_j is part of the model $B_j \in M(S_i)$. The dictionary-texts T_i^D are grouped into groups that have the same basic-form $T_i^D \in T_i^D(B_j) if \beta_j(T_i^D) = B_j$ and each group of tasks $T_i^D(B_j)$ is further grouped into groups that have the same query-form $T_i^D \in T_i^D(B_j(\theta_k)) if \Theta_k(T_i^D(\beta_j)) = \theta_k$. The query-form $\Theta_k(T_i^D)$ is learned for each group of tasks $T_i^D(B_j(\theta_k))$ by transforming the dictionary-text into a set of coding categories $\Theta_k(T_i^D) \equiv \theta_k = \{c_k^1,...,c_k^N\}$, where N is a number of coding categories k-th query-form θ_k. For the dictionary-text T_i^D, the words of the text $w_i \in T_i^D$ are coded by applying the coding categories c_i stored in the categorexicon $c_i \in C$. As the result of learning the query-forms, that consist of the coding categories $\theta_k = \{c_k^1,...,c_k^N\}$, are obtained.

During understanding process the learned query-forms θ_k are used to find the basic-form, the procedural-form and the interpretational script. During understanding process an examined text T^0 is transformed into a set of coding categories $t = \{k^1,...,k^M\}$, where M is a number of coding categories. At the first level of understanding, the process of understanding is described by the following algorithms:

FOR k=1 to K if $t \equiv \theta_k$ then apply $[B_k, P_k, S_k]$, END.

In the case when forms $[B_k, P_k, S_k]$ are not found, this is an indication that SUS can not understand the dictionary-text. The query-form θ_k is implemented in the form of the clause $if((k_i^1 \in \theta_k) \wedge (k_i^2 \in \theta_k),...,\wedge(k_i^M \in \theta_k)) than E(B_k)$, where $E[B_k]$ denotes the basic-form.

In this Section, learning of the mineralogical-dictionary-text $T^D[Min]$ as an example of learning of the dictionary-text is described. The mineralogical-dictionary-text $T^D[Min]$ is a short text that could be found in mineralogical books, journals, encyclopedias, dictionaries or magazines it has specific form and includes categories of the mineralogical knowledge. Examples of the mineralogical-dictionary-texts used for learning of the dictionary-text are presented in Section 7.3.3. The basic mineralogical knowledge is acquired from the mineralogical books as described in Section 7.3.3. The knowledge given in a formal style such as in the TEXT 1 is learned by transforming the data into the basic-form specified by the mineral model $M[Min]$. The complex data, needed for naming of the object from the mineral category, are acquired from the different available sources and stored in the text format in the Mineral-Sensory-Expert. The Mineral-Sensory-Expert uses stored knowledge during naming of the minerals. Example of data used for learning of the basic-form are as follows:

Name: Acanthite ChemicalFormula: Ag2S EmpiricalFormula: Ag2S Density: 7.200000 Hardness: 2.000000 Color: Lead gray, Gray, Iron black Streak: shining black Cleavage: [001] Poor, [110] Poor Luster: Metallic Diaphaneity: Opaque Fracture: Sectile Habit: Arborescent , Blocky , Skeletal Luminescence: None Magnetism: Nonmagnetic Radioactivity: Not Radioactive.

Acquired mineralogical data can contain some errors and need to be corrected during incremental learning process. The mineralogical data are part of the mineralogical knowledge. The learned mineralogical knowledge is used for understanding of the sensory object and for understanding of the mineralogical-dictionary-text $T^D[Min]$. During understanding of the sensory object the learned knowledge implemented as part of the Mineral-Sensory-Expert is used to name an examined object. Learning of the knowledge of the sensory object is described in Section 7.2. An object that is named can be further interpreted by accessing knowledge stored in the Mineral-Interpretational-Script-Expert. The Mineral-Interpretational-Script-Expert stores the basic-forms that are extracted from mineralogical-dictionary-texts. The basic-forms are used to define the basic meaning of the mineralogical-dictionary-text. These basic-forms implemented as a part of the Mineral-Interpretational-Script-Expert are invoked as a method of FindBasicMeaning(Q) during understanding of the mineralogical-dictionary-text.

The knowledge implementation, the term that is introduced in this book, stress two aspects of SUS learning; namely the learning of the knowledge and learning of the skills. The important part of the knowledge implementation is checking the correctness and the consistency of knowledge. The Mineral-Dictionary-Script-Expert stores knowledge that is acquired from different sources. Knowledge that is acquired from different sources can be incomplete and even incorrect. The mineralogical data that are part of the mineralogical knowledge refers to the deferent attributes of the mineral category. The values of the minerals attributes are stored as real numbers, or as the categorical data. The values of the attribute that are acquired from the different sources are compared and if there is no significant difference among them the proper value can be easily selected. In the case of differences in the values of the attribute the consultation with a mineralogical expert is needed. In the case of the categorical data such as colour the problem with selection of the proper value can not be easily solved. The data such as colour are more subjective so the different experts can give the different advice that is based on their own preferences. In order to avoid an error, the basic colour categories are established and the color categories gathered from sources such as TEXT 1 are transformed into the basic colour categories, as described in Section 7.3.3.

As it was described in Section 7.3.3, the basic meaning of the mineralogical-dictionary-text is given in terms of the name, the chemical composition, the crystal structure, and the characteristic and additional features. The query-form is learned based on the previously learned coding categories. The coding categories are established based on the 'minerals model'. Here are a few examples of the learned coding categories: CMineralName {andradite, atacamite,...}, CMineralAttr {hardness, lustre, special gravity, ...}, CValColour {yellow, yellow green,...}. The query-form is learned as the result of analysis and generalization of the mineralogical-dictionary-texts. For example, the mineralogical-dictionary-text "Vanadinite is transparent to translucent mineral and it has resinous to subadamantine lustre" is transformed into the query-form "CMineralName CAuxIs CValDiaphaneity CAuxAnd CAuxHas CMineralAttr CValLustre" by applying learned coding categories.

The basic-form of the mineralogical-dictionary-texts is learned based on the query-form. In comparison to the text-task the dictionary-text need to be learned at more specific categorical level in order to find exact meaning. Each basic meaning is given by the basic-form and is related to the interpretational script. As it was previously described, the interpretational script $S(T^D)$ is given by the model of the phenomena $M(S_i)$ to which the script refers. The model of the mineral object is described in Section 7.3.3. The model include the physical attributes a_j^{Ph} such as the size $a(L)_j^{Ph}$, the specific gravity $a(G)_j^{Ph}$, the hardness $a(H)_j^{Ph}$, the color $a(C)_j^{Ph}$, the streak $a(S)_j^{Ph}$, the cleavage $a(K)_j^{Ph}$, the diaphaneity $a(D)_j^{Ph}$, the luster $a(L)_j^{Ph}$, the habit $a(H)_j^{Ph}$, the tenacity $a(T)_j^{Ph}$, the fluorescence $a(F)_j^{Ph}$, the magnetism $a(M)_j^{Ph}$, the radioactivity $a(R)_j^{Ph}$, the piezoelectricity $a(P)_j^{Ph}$, the chemical attribute a_j^{Ch}, the optical attributes a_i^{P} and the X Ray diffraction analysis attributes a_i^{X}.

All minerals have distinct crystal structures given by two characteristic attributes $s_i\left[s_i^Z, s_i^W\right]$. The symmetry attribute s_i^W is expressed by the symbol for a space group that consists of the centering followed by an abbreviated Hermann-Mauguin symbol, whereas the cell attribute parameters s_i^Z is represented by the axial ratios $s(a)_i^Z$, the cell dimensions $s(d)_i^Z$, the unit cell $s(u)_i^Z$ and the volume $s(v)_i^Z$. Each mineral is given by its name χ_i and the special index ∂_i. The name χ_i is expressed in one of existing language L_j such as English $\chi_i^N(E)$, or Polish $\chi_i^N(P)$. The name is also connected with the synonymies that are given in one of the known language such as English $\chi_i^S(E)$. The name can be written as $\chi_i[\chi_i^N(L_j), \chi_i^S(L_j)]$. The special index ∂_i refers to one of the systems of Nickel-Strunz, symbol ∂_i^{N-S}, Hey's Index ∂_i^H or Dana index ∂_i^D and can be written as $\partial_i[\partial_i^{N-S}, \partial_i^D, \partial_i^H]$. The knowledge of the locality λ_i where the mineral occur is very useful information in process of recognition λ_i. The locality is referring to the geographical knowledge and usually includes the country λ_i^C, the province λ_i^P, the place λ_i^L, the mine λ_i^M and is expressed as $\lambda_i[\lambda_i^C, \lambda_i^P, \lambda_i^L, \lambda_i^M]$. Example of the description of the locality is as follows: White Queen Mine, Hiriart Mountain, Pala District, San Diego Co., California, USA.

The mineral category is part of the rock category $\ldots \rhd \left\langle \nu_{Rock} \right\rangle \succ [\nu_{Min}]$ and the part of the mineralogical knowledge is concerned with the geological setting γ_i and the associated minerals θ_i. The name of origin Λ_i is given by the year of naming Λ_i^Y, the name of the discoverer Λ_i^N, the name of the source Λ_i^S and can be written as $\Lambda_i[\Lambda_i^Y,\Lambda_i^N,\Lambda_i^S]$. Mineral can be named Γ_i for name of the professional Γ_i^N, the appearance Γ_i^A and can be written as $\Gamma_i[\Gamma_i^N,\Gamma_i^A]$. The general basic-form of the mineralogical dictionary-text is given in the form of the mineralogical model that is part of the interpretational script, as follows:

$$\nu_M^i \equiv \varphi_i[\phi_i^{ChC},\phi_i^F,\phi_i^T]s_i[s_i^Z,s_i^W]a_i[d_j^{Ph},d_j^{Ch},d_j^P,d_j^X]\chi_i[\chi_i^N(L),\chi_i^S(L)]\partial_i[\partial_i^{N-S},\partial_i^P,\partial_i^H]\hbar_i[\hbar_i^Y]\lambda_i[\lambda_i^C,\lambda_i^P,\lambda_i^L,\lambda_i^M]\Lambda_i[\Lambda_i^Y,\Lambda_i^N,\Lambda_i^S]\Gamma_i[\Gamma_i^N,\Gamma_i^A]\gamma_i\theta_i .$$

The specific basic meaning of the dictionary-text is concerned with meaning of the one line (sentence) of the dictionary-text. The specific basic meaning is represented in the formal style and implemented as the part of the Mineral-Dictionary-Script-Expert. For example, the basic meaning represented in the formal style $val(a(C)_j^{Ph}) \equiv \{c_1^C,\ldots,c_N^C\}(\nu)$ as the basic-form is stored as the method ImpMark_Colour_Is_() that is invoked by the Mineral-Dictionary-Script-Expert.

During learning of the query-form the mineralogical-dictionary-texts are grouped into groups that have the same meaning. For example, for the group of dictionary-texts :"gold is opaque and its luster is metallic", "bismuth is opaque with metallic luster", "Vanadinite is transparent to translucent mineral and it has resinous to subadamantine luster" the query-form is implemented as a clause "CAT_IS"; "CAT_VALDIAPHANEITY"; "CAT_VALLUSTER"; "CAT_LUSTER"; as shown in the listed C++ program. The query-form consists of coding categories such as "CAT_IS", "CAT_VALDIAPHANEITY"; "CAT_VALLUSTER"; "CAT_LUSTER". Learned query-form is stored in an array m_StringZDANIE. Understanding of the mineralogical-dictionary-texts $T^D[Min]$ requires transforming it into the query-form and into the interpretational script J. Understanding text such as "gold is opaque and its luster is metallic" is to compare meanings of this text with the meaning of a basic-form that is part of the interpretational script. The basic-form for these texts is given as: "Minera-Name has ValLuster and ValDiaphaneity". The method Compare() evaluates an examined dictionary-text and invokes the Mineral-Knowledge-Expert to find its meaning and to understand the text. The arrays m_ArCatMineralDiaphaneity and m_ArCatMineralLuster store values of the attribute lustre and diaphaneity. The Evaluate() method evaluates if the mineral gold in the text "gold is opaque and its luster is metallic" has values that are stored in arrays m_ArCatMineralDiaphaneity and m_ArCatMineralLuster. The result of the basic understanding process is stored in the form of the string m_Explanation.

```
m_StringZDANIE[0]="CAT_IS";
m_StringZDANIE[1]="CAT_VALDIAPHANEITY";
m_StringZDANIE[2]="CAT_VALLUSTER";
m_StringZDANIE[3]="CAT_LUSTER";
m_NoZdanie=4;m_NoTruth=0;
Compare();
if(m_NoTruth>=m_NoZdanie)
{pMineralKnowledgeExpert->Mineral_dafaneity_luster(&m_INPUT);}

void CMineralKnowledgeExpert::Mineral_dafaneity_luster(CString
*StringZbiorczy)
{
Evaluate(&m_StringZbiorczy);
if(m_OCatMineralName=="*")
{m_Explanation="It";}
Else
{m_Explanation=m_OCatMineralName;}
 m_Explanation+=" has diaphaneity";
for(j=0;j<m_NoArCatMineralDiaphaneity;j++)
{m_Explanation+=m_ArCatMineralDiaphaneity[j];m_Explanation+=" ";}
 m_Explanation+=" has luster ";
for(j=0;j<m_NoArCatMineralLuster;j++)
{m_Explanation+=m_ArCatMineralLuster[j];}

}
```

The listed C++ program shows implementation of the method Mineral_Hardness_SpecificGravity of the second level of understanding. Understanding at this level, in addition to comparison of the values of the attributes of the mineral described in the examined dictionary-text, explain the text in terms of values of mineral attributes that are part of the interpretational script. The method Mineral_Hardness_SpecificGravity is searching through the interpretational script represented by procedures Hardness_Mohs_Category () or SpecificGravity_Number_Category to find additional information about the mineral presented in the examined dictionary-text. The additional information in the form of values of attributes such as hardness or specific gravity is used in reasoning process or explanatory process. The procedure MeasurementUnit selects the proper units of the attribute that is selected for explanatory process. Procedures Definition() and Def_Mineral_Is() supply an information about selected attribute in the form of the short definition. Procedures Select_Name(), HabitCoding(), ColorCoding() and DecodingCategories() decode values of attributes color and habit and prepare the data for the final result of the explanatory process.

```
voidCMineralKnowledgeExpert::Mineral_Hardness_SpecificGravity(CString
*StringZbiorczy)
   {
   POMOC(&m_StringZbiorczy);
   Hardness_Mohs_Category((int)m_No1);
   m_Explanation+=m_Hardness_MohsCategory;
   SpecificGravity_Number_Category(m_No1);
   m_Explanation+=m_SG_Nomber_Category;
   MeasurementUnit();
   m_Explanation+=m_ResultConvertion[0];
   Definition();
   Def_Mineral_Is();
   Select_Name();
   HabitCoding();
   ColorCoding();
   DecodingCategories();
   }
```

The first level of understanding of the dictionary-text is to compare the procedural-form of the examined text with the basic-form of the model. If the procedural-form of the examined text is "equal" the procedural-form of the model $P(T) \equiv P(M)$ the text is understood at the first level of understanding. The procedural-form of the dictionary-text is the "updated" version of the basic-form that means the specific data are accessible from the procedural-form. On the learned basic-form, understanding process is founded. All linguistic description that is derived from a given basic-form can be learned as a text form that needs to be translated into the basic-form.

Basic-forms are learned from the selected mineral-dictionary-texts. The basic-form is represented in the formal style that is easy to implement and can be easy to understand. For the dictionary-text the basic-form is closely related to the query-form and is used to learn the query-forms. The basic-forms are represented by formal definitions that are used to build the model of the interpretational script.

Examples of the basic-forms used to learn to understand the mineralogical-dictionary-text are shown below. These basic-forms are given with the examples of the dictionary-texts from which they were extracted. In this context understanding of the symbols used in description of the basic-forms should be easy to follow. Examples of the basic-forms are as follows:

"Colour varies from greenish to golden yellow to brown"; ="Colour ranges from bright red and orange red to brownish red brown or yellow"; "Colour is greenish yellow to yellow"; "mineral is reddish silver white to grayish white"; "cobaltite colour varies from grayish black to reddish silver white"; "hauerite is reddish-brown to dark-brown in colour"; "It is yellow or greenish yellow"; "Rarely it can be brown yellow reddish brown or black";

$$val(a(C)_j^{Ph}) \equiv \{c_1^C,...,c_N^C\}(v)$$

"It is silvery white with reddish or iridescent tarnish"; "cobaltite has colour steel-grey often with violet-to-purple tinge"; "cobaltite colour range from steel-grey to silver-white often with reddish tinge" "Bright rich yellow colour is re-sistant to tarnishing"

$$val(a(C)_j^{Ph}) \equiv \{c_1^C,...,c_N^C\}(\nu)\chi\{c_1^\chi,...,c_N^\chi\}(\nu)$$

"This lustre is not altered by tarnishing if mineral is exposed to atmosphere"
"Cuprite can turn superficially dark grey on exposure to light", "Silver is silvery white in colour thought it tarnishes on exposure to atmosphere";

$$val(a(L)_j^{Ph}) \equiv \{c_1^L,...,c_N^L\}(\nu)\chi\{c_1^\chi,...,c_N^\chi\}(\nu)A\ (E)\{c_1^E,...,c_N^E\}$$

"cobaltite colour range from steel-grey to silver-white often with reddish tinge and its crystals are pink";

$$val(a(C)_j^{Ph}) \equiv \{c_1^C,...,c_N^C\}(\nu)val(a(Ck)_j^{Ph}) \equiv \{c_1^{Ck},...,c_N^{Ck}\}(\nu)\chi\{c_1^\chi,...,c_N^\chi\}(\nu)$$

"Crystals form as cubes or octahedral but are rare"; Crystals take form of cubes but are uncommon", "It is rare for copper to form crystals when it does they take form of cubes octahedral or dodecahedral";

$$val(a(Hc)_j^{Ph}) \equiv \{c_1^{Hc},...,c_N^{Hc}\}(\nu)$$

"fergusonite-(Y) crystals are prismatic to pyramidal and it is black to brownish-black in colour";

$$val(a(Hf)_j^{Ph}) \equiv \{c_1^{Hf},...,c_N^{Hf}\}(\nu)val(a(C)_j^{Ph}) \equiv \{c_1^C,...,c_N^C\}$$

"zinc aluminum oxide gahnite is member of spinel group forming octahedral crystals";

$$\varphi_j^N \aleph_j\{c_1^\aleph,...,c_N^\aleph\}val(a(Hc)_j^{Ph}) \equiv \{c_1^{Hc},...,c_N^{Hc}\}.$$

Similarly the definition-mineralogical-dictionary script that refers to the definition of the mineralogical concepts is transformed into the basic-form. The example of the definition of the attribute "hardness", which shows the different aspects of the defined concept, is given to explain the complexity of learning the general mineralogical knowledge. Each general concept can be defined in different way by referring to other concepts that need to be learned to make it possible to understand the text. However, there will be always some mineralogical concepts that are not learned yet, and concepts that belong to other scientific domains such as physics or chemistry. In this example, the concepts "abrasion" or "scratching" are not typical mineralogical concepts. Also, the concept of Mohs scale needs to be previously learned. Examples of the dictionary-text-tasks are as follows:

Hardness (H) is [measure of] [resistance of a mineral] [to scratching];
Hardness (H) is [measure of] [how easy mineral] [can be scratched];
Hardness is [property of] [resisting abrasion or scratching];
Hardness is [attribute of mineral] [resisting abrasion or scratching].

Learning of the dictionary-text-tasks will be the topic of further research.

References

1. Les, Z., Les, M.: Shape Understanding System. The first steps toward the visual thinking machines. Springer, Berlin (2008)
2. Matthews, E.: Mind. In: Mullarkey, J., Williams, C. (eds.) Key Concepts in Philosophy. Continuum, London (2005)
3. Feser, E.: Philosophy of Mind. OneWorld, Oxford (2010)
4. Polesel, J., Dulfer, N., Turnbull, M.: The experience of education: the impacts of high stakes testing on school students and their families. Literature review, pp. 1–15. Withlam Institute, Sydney (2012)
5. Alexander, R.J. (ed.): Children, their World, their education: final report of the Cambridge Primary Review. Routledge, London (2010)
6. Dawson, C.: The crisis of western education. The Catholic University of America Press (Sheed and Ward), New York (1961)
7. Wolf, J.G.: Cognitive development as optimization. In: Bolc, L. (ed.) Computational Model of Learning. Springer, Berlin (1987)
8. Howard, R.W.: Concepts and Schemata: an introduction. Cassel Educational, Philadelpia (1987)
9. Gennari, J., Langely, P., Fisher, D.: Models of incremental concept formation. Artificial Intelligence 40, 11–61 (1989)
10. Gilson, E.: The Spirit of Mediaeval Philosophy. University of Notre Dame Press, London (1991)
11. Rosh, E., Mervis, C.B.: Family Resemblance: Studies in the Internal Structure of Categories. Cognitive Psychology 7, 537–605 (1975)
12. Anderson, J.R.: The architecture of cognition. Harvard University Press, Massachutes (1983)
13. Piaget, J.: The Origins of Intelligence in Children. Routledge and Kegan Paul, London (1953)
14. Arnheim, R.: Visual Thinking. Faber and Faber, London (1970)
15. Gibson, J.J.: The Perception of the Visual World. Houghton, Boston (1950)
16. Marr, D.: Vision. Freeman, San Francisco (1982)
17. Guthrie, E.R.: The Psychology of Learning. Harper, New York (1935)
18. Hull, C.L.: Mathematico-Deductive Theory of Rote Learning. Greenwood Pub Group, London (1940)
19. Tolman, E.C.: Behavior and psychological man: essays in motivation and learning. Univ. of California Press, Berkeley (1951)
20. Kohen, M.: Representations and algorithms for cognitive learning. Artificial Intelligence 5, 199–216 (1974)
21. Les, M., Murphy, M.: Computer based assessments - does prior computer competence counts? In: The AARE International Education Research Conference (paper presented), Brisbane (2002)

22. Australian Curriculum, Home page (2011), http://www.acara.edu.au/default.asp
23. Les, M.: The Influence of Computerisation on Changes on the Didactic Situation. In: Moroz, H. (ed.) Didactical Situations. Silesian University, Katowice (1989)
24. English and Mathematics. Excel basic skills. Pascal Press, Glebe NSw (2011)
25. 501 Algebra questions. Learning Express, New York (2011)
26. Foresti, G.L., Fabris, F.: Performance Evaluation of a Multisensor System for Autonomus Guidance Vehicle. In: International Conference on Automotive and Transportation Technology, ISATA, Dublin (2000)
27. Cheok, K.A., Smid, G.E., Warner, C.: Evaluation of Multi-Sensors and AI Approaches for Collision Avoidance. In: International Conference on Automative and Transportation Technology, ISATA, Vienna (1999)
28. Friedrich, H., Rogalla, O., Dillmann, R.: Communication and Propagation of Action Knowledge in Multi-agent Systems. Robotics and Autonomous Systems 28(4), 290–295 (1999)
29. Kumar, R.A., Menon, A.: Collision Avoidance in a Multi-Robot System by Emulating Human Behaviour. In: International Conference on Information Systems Analysis and Synthesis and World Multiconference on Systemics, Cybernetics and Informatics, Orlando (2001)
30. Brooks, R.A., Breazeal, C., Marjanovic, M., Scassellati, B., Williamson, M.M.: The cog project: Building a humanoid robot. In: Nehaniv, C.L. (ed.) CMAA 1998. LNCS (LNAI), vol. 1562, pp. 52–87. Springer, Heidelberg (1999)
31. Weng, J., Hwang, W.S., Zhang, Y., Yang, C., Smith, R.: Developmental Humanoids: Humanoids that Develop Skills Automatically. In: IEEE-RAS International Conference on Humanoid Robots, Cambridge (2000)
32. Buchanan, B.G., Barstow, D., Bechtal, R., Bennet, J., Clancey, W., Kulikowski, C., Mitchell, T., Waterman, D.A.: Constructing an Expert System. In: Hayes-Roth, F., Waterman, D.A., Lenat, D.B. (eds.) Building Expert System, pp. 68–127. Addison-Wesley, Massachusetts (1983)
33. Johnson, P.E.: What kind of expert should a system be? Journal of Medical Philosophy 8, 77–97 (1983)
34. Buchanan, B.G., Shortlife, E.H.: Rule-Based Expert System. The MYCIN Experiments of the Stanford Heuristic. Programming Project. Addison-Wesley Publishing Company, New York (1984)
35. Hwee, J., Building, L.: Visual Vocabulary for Image Indexation and Query Formulation. Pattern Analysis & Applications 4(2-3), 125–139 (2000)
36. Fournier, J., Cord, M., Philipp-Foliguet, S.: RETIN: A Content-Based Image Indexing and Retrieval System. Pattern Analysis & Applications 4(2-3), 153–173 (2001)
37. Chaudhuri, B.B., Garain, U.: An Approach for Recognition and Interpretation of Mathematical Expressions in Printed Document. Pattern Analysis & Applications 3(2), 120–131 (2000)
38. Kim, J.H., Kim, K.K., Suen, C.Y.: HMM-MLP Hybrid Model for Cursive Script Recognition. Pattern Analysis & Applications 3(4), 314–324 (2000)
39. Henning, A., Sherkat, N.: Cursive Script Recognition using Wildcards and Multiple Experts Pattern. Analysis & Applications 4(1), 289–302 (2001)
40. Koerich, A.L., Sabourin, R., Suen, C.Y.: Recognition and verification of unconstrained handwritten words. IEEE Transactions on Pattern Analysis and Machine Intelligence 27(10), 1509–1522 (2005)

41. Koerich, A.L., Sabourin, R., Suen, C.Y.: Large vocabulary off-line handwriting recognition: A survey. Pattern Analysis & Applications 6(2), 97–121 (2003)
42. Tolba, A.S., Abu-Rezq, A.N.: Combined Classifiers for Invariant Face Recognition. Pattern Analysis & Applications 3(4), 289–302 (2000)
43. Ablameyko, S., Gorelik, A., Medvedev, S.: Reconstruction of 3D Object Models from Vectorised Engineering Drawings Formulation. Pattern Analysis & Applications 5(1), 2–14 (2002)
44. Quinlan, J.R.: Induction of Decision Trees. Machine Learning 1, 81–106 (1986)
45. Breiman, L., Friedman, J.H., Olshen, R.A., Stone, C.J.: Classification and Regression Tree. Wadsworth International Group, California (1984)
46. Michalski, R.: Theory and Methodology of Inductive Learning. Artificial Intelligence 20, 111–161 (1983)
47. Minton, S., Carbonell, J.G., Knoblock, C.A., Kuokka, D.R., Etzioni, O., Gil, Y.: Explanation-Based Learning: A problem Solving Perspective. In: Carbonell, J.G. (ed.) Machine Learning: Paradigms and Methods, Cambridge, Massachusetts, pp. 63–118 (1990)
48. Sestito, S., Dillon, T.S.: Automated Knowledge Acquisition. Prentice Hall, New York (1994)
49. Michalski, R.S., Carbonell, J.G., Mitchel, T.M. (eds.): Machine learning: An Artificial Intelligence Approach, vol. 1. Tioga Publishing, San Mateo (1983)
50. Michalski, R.S., Carbonell, J.G., Mitchel, T.M. (eds.): Machine learning: An Artificial Intelligence Approach. Morgan Kaufmann, San Mateo (1986)
51. Zurada, J.M.: Introduction to Artificial Neural Systems. West Publishing Company, Minnesota (1992)
52. Shank, R.C.: Conceptual dependency: A theory of natural language understanding. Cognitive Psychology 3, 552–631 (1972)
53. Pearl, J.: Probabilistic Reasoning in Intelligent Systems: Network of Plausible Inference. Morgan Kaufman, San Francisco (1988)
54. Isham, V.: An Introduction to Spatial Processes and Markov random field. International Statistic Reviev 49, 21–43 (1981)
55. Rich, E., Knight, K.: Artificial Intelligence. McGraw-Hill, New York (1991)
56. Kohonen, T.: Self-organization systems. Springer, Berlin (1984)
57. Alvarado, S.J.: Understanding Editorial Text: A computer Model of Argument Comprehension. Kluver Academic Publishers, Boston (1991)
58. Dnis, M.: Image and Cognition. Harvester, New York (1989)
59. Bramer, M.A.: Research and Development in Expert Systems III. British Informatics Society Ltd., London (1987)
60. Brownston, L., Forrel, R., Kant, E., Martin, N.: Programming Expert Systems in OPS5. An Introduction to Rule-Based Programming. Adison-Wesley Publishing Company, Massachusetts (1985)
61. Gupta, M., Kandel, A., Bandler, W., Kiszka, J.B.: Approximate Reasoning in Expert Systems. North-Holland, Amsterdam (1988)
62. Negoita, C.V.: Experts Systems and Fuzzy Systems. Cummings Publishing Company, Menlo Park (1985)
63. Les, Z.: Objects Recognition System, vol. 60, pp. 33–48. Bulletin of St. Staszic University of Mining and Metallurgy, Cracow (1992)
64. Kowalski, R.: Logic for Problem Solving. North-Holland, Oxford (1979)
65. Grabowski, J., Lesconne, P., Wechler, W.: Algebraic and Logic Programming. Akademie-Verlag, Berlin (1988)

66. Bratko, I., Mugletton, S.: Application of Inductive Logic Programming. Communications ACM 38(11), 65–70 (1995)
67. Jensen, F.V.: An Introduction to Bayesian Networks. Springer, New York (1996)
68. Muller, J.P.: Control Architecture for Autonomous and Interacting Agents: A Survey. In: Cavedon, L., Rao, A., Wobcke, W. (eds.) Intelligent Agent Systems. Springer, Berlin (1996)
69. Leake, D.B. (ed.): Case-Based Reasoning. Experience, Lessons and Future Directions. AAAI Press/MIT Press, Menlo Park (1996)
70. Binford, T.O.: Survey of Model-Based Image Analysis Systems. International Journal of Robotics Research 1(1), 18–64 (1982)
71. Matsuyama, T., Hwang, V.: SIGMA: a Knowledge-based Aerial Image Understanding System. Plenum Press, New York (1990)
72. Ullman, S., Richards, W.: Image Understanding. Ablex Publishing Corporation, Norwood (1989)
73. Shirai, Y.: Three-Dimensional Computer Vision. Springer, Berlin (1987)
74. Overington, I.: Computer Vision. Elsevier Science Publisher, Amsterdam (1992)
75. Wechsler, H.: Computational Vision. Academic Press, London (1990)
76. Hanson, A., Riseman, E.: Visions: a computer system for interpreting scenes. In: Hanson, A., Riseman, E. (eds.) Computer Vision Systems, pp. 303–333. Academic Press, New York (1978)
77. Tsotsos, J.K., Mylopoulos, J., Covvey, H.D., Zucker, S.W.: A framework for visual motion understanding. IEEE Transactions on Pattern Analysis and Machine Intelligence 2(6), 563–573 (1980)
78. Brooks, R.: Symbolic Reasoning Among 3-Dimensional Models and 2-Dimensional Images. Artificial Intelligence 17, 285–349 (1981)
79. Draper, B.A., Collins, R.T., Brolio, J., Hanson, A.R., Riseman, E.M.: The schema system. International Journal of Computer Vision 2, 209–250 (1989)
80. Dillon, C., Caelli, T.: Cite - scene understanding and object recognition. In: Caelli, T. (ed.) Machine Learning and Scene Interpretation, pp. 119–187. Plenum Publishing Cooperation, New York (1997)
81. Dance, S., Caelli, T.: Soo-pin: Picture Interpretation networks. In: Caelli, T., Bishof, W.F. (eds.) Advances in Computer Vision and Machine Intelligence, pp. 225–254. Plenum Publishing Cooperation, New York (1997)
82. Salzbrunn, R.: Wissensbasierte Erkennung und Lokarisierung von objekten. Shaker Verlag, Aachen (1997)
83. Tadeusiewicz, R., Ogiela, M.R.: Medical Image Understanding Technology. Springer, Berlin (2004)
84. Bhanu, B., Faugeras, O.D.: Shape Matching of Two-Dimensional Objects. IEEE Transactions on Pattern Analysis and Machine Intelligence 6(2), 137–156 (1984)
85. Lu, C.H., Dunham, J.G.: Shape Matching Using Polygon Approximation and Dynamic Alignment. Pattern Recognition Letters 14, 945–949 (1993)
86. Pavlidis, T., Ali, F.: A Hierarchical Syntactic Shape Analyzer. IEEE Transactions on Pattern Analysis and Machine Intelligence PAMI-1(1), 2–9 (1977)
87. You, K.C., Fu, K.S.: Distorted Shape Recognition Using Attributed Grammars and Error-Correcting Techniques. Computer Graphics and Image Processing 13, 1–16 (1980)
88. Bala, J., Wechsler, H.: Shape Analysis Using Genetic Algorithms. Pattern Recognition Letters 14, 965–973 (1993)

89. Bala, J., Wechsler, H.: Shape Analysis Using Hybrid Learning. Pattern Recognition 29(8), 1323–1333 (1996)
90. He, Y., Kundu, A.: 2-D Shape Classification Using Hidden Markov Model. IEEE Transactions on Pattern Analysis and Machine Intelligence 13(11), 1172–1184 (1991)
91. Kartikeayan, B., Sarkar, A.: Shape Description by Time Series. IEEE Transactions on Pattern Analysis and Machine Intelligence 11(9), 977–984 (1989)
92. Pal, N.R., Pal, P., Basu, A.K.: A New Shape Representation Scheme and Its Application to Shape Discrimination Using a Neural Network. Pattern Recognition 26(4), 543–551 (1993)
93. Mahmoud, S.I.: Arabic Character Recognition Using Fourier Descriptors and Character Contour Encoding. Pattern Recognition 27(6), 815–824 (1994)
94. Samal, A., Edwards, J.: Generalized Hough Transform for Natural Shapes. Pattern Recognition Letters 18, 473–480 (1997)
95. Besl, P.J., Jain, R.C.: Three-dimensional Object Recognition. Computing Surveys 17(1), 75–154 (1985)
96. Chin, R.T., Dyer, C.R.: Model-Based Recognition in Robot Vision. Computing Surveys 18(1), 67–108 (1986)
97. Pope, A.R.: Model-based Object Recognition, a Survey of Recent Research. Department of Computer Science, The University of British Columbia (1994)
98. Locke, J.: An Essay Concerning Human Understanding. In: Yolton, J. (ed.) Dent, London (1961)
99. Hume, D.: Enquires Concerning Human Understanding and Concerning the Principles of Morals. In: Selby-Bigge, B. (ed.) Oxford University Press, Oxford (1975)
100. Berkeley, G.: Principles of Human Knowledge and Three Dialogues. In: Robinson, H. (ed.) The World's Classics. Oxford University Press, Oxford (1996)
101. Leibnitz, G.W.: New Essay on Human Understanding. In: Remnant, P., Bennet, J. (eds.) Cambridge Texts in the History of Philosophy. Cambridge University Press, Glasgow (1996)
102. Kant, I.: Critique of Pure Reason. In: Politis, V. (ed.) The Everyman Library. Everyman, London (1996)
103. Popper, K., Eccles, J.C.: The Self and its Brain. Routledge, London (1977)
104. Rock, I.: The Logic of Perception. The MIT Press, Cambridge (1983)
105. Hebb, D.O.: The organization of behavior. Willey, New York (1949)
106. Les, Z.: An Aesthetic Evaluation Method Based on Image Understanding Approach. In: The First International Conference on Visual Information Systems VISUAL 1996, Melbourne (1996)
107. Les, Z.: Image Understanding and Aesthetic Evaluation, vol. electronic version. QJF Press, Melbourne (2010)
108. Parker, J.R.: Algorithms for image processing and computer vision. John Wiley & Sons, Inc., New York (1997)
109. Les, Z., Les, M.: Shape Understanding System - Learning to Solve Text Tasks. International Journal of Understanding (electronic version) 1(1), 60–97 (2009)
110. Les, M., Les, Z.: Problem Solving, Statistical Analysis and Visualisation. In: ASCILITE 1995 Conference, Melbourne (1995)
111. Colaruso, R., Hammil, D.: Motor Free Visual Perception Test. Academic Therapy Publications, New York (2003)
112. Gardener, M.F.: Test of Visual-Perceptual Skills. Psychological and Educational Publications (1996)
113. Roid, G.: Stanford-Binet Intelligence Scale. Riverside, Chicago (2003)

114. Gluting, J., Adams, W., Shwslow, D.: Wide Range Intelligence Test. Wide Range, New York (1999)
115. Sala, D.S., Gray, C., Baddeley, A., Wilson, L.: Visual Patterns Test. Thames Valley Test Company, New York (1997)
116. Meyers, J.E., Meyers, K.R.: Rey Complex Figure Test and Recognition Trial. PAR, Boston (1996)
117. Huber, H.: Rorschach Psychodiagnostic Test. Hans Huber, Berne (1948)
118. Haiman, J.: Dictionaries and encyclopedias. Lingua 50, 329–357 (1980)
119. Athanasou, J., Deftereos, A.: Year 9 NAPLAN-style Tests. Pascal Press, Sydney (2010)
120. Pellant, C.: Rocks and minerals. Dorling Kindersley Limited, London (1992)
121. Mottana, A., Crespi, R., Liborio, G.: Guide to rocks and minerals. Simon & Shuster Inc., New York (1977)
122. Bonewitz, R.L.: Rock and Gem. Dorling Kindersley Limited, London (2005)
123. Britannica, Deluxe Edition CD-ROM (2001)
124. Rogerson, D.: Inside COM. Microsoft Press, Redmond (1997)
125. Les, Z.: Shape Understanding. Possible Classes of Shapes. International Journal of Shape Modelling 7(1), 75–109 (2001)
126. Les, Z., Les, M.: System of Experts to Perform an Epistemological Task of Shape Understanding. In: The IASTED International Conference on Software Engineering, Las Vegas (2000)
127. Les, Z., Les, M.: Shape Understanding as Knowledge Generation. Transaction on Data and Knowledge Engineering 16(3), 343–353 (2004)
128. Les, Z., Les, M.: Shape Understanding System: Categorical Learning. International Journal of Understanding (electronic version) 2(1), 1–50 (2011)
129. Geeraerts, D.: Introduction: Prospects and problems of pro-totype theory. Linguistics 27, 587–612 (1989)
130. O'Connel, M., Airey, A.: The complete encyclopedia of signs and symbols. Hermes House, New York (2010)
131. Brittanica 2001. Oxford University Press, Oxford (2001)
132. Visual Dictionary. DK, London (2006)
133. Horton, W.: The Icon Book. Visual Symbols for Computer System and Documentation. Jon Wiley & Sons, Inc., New York (1994)
134. Bliss, C.K.: Semantography (Blissymbolics). Semantography (Blissymbolics) Publication, Sydney (1963)
135. Lunde, P. (ed.): The secrets of code. Weldon Owen Inc., San Francisco (2009)
136. Les, Z.: The Use of Prolog as a Language for Creation an Intelligent Recognition Systems, vol. 60. Automatics, Scientific Bulletins of St. Staszic University of Mining and Metallurgy, Cracow (1992)
137. Kay, P., McDaniel, C.K.: The linguistic significance of the meanings of basic color terms. Language 54(3), 610–646 (1978)
138. Beiser, A.: Shaum' Outline of Theory and Problems of Applied Physics. McGraw-Hill, New York (2004)
139. Bueche, F.J., Hecht, E.: Shaum' Outline of Theory and Problems of College Physics. McGraw-Hill, New York (2006)
140. Shmidt, P.A., Ayers, F.: Shaum' Outline of Theory and Problems of College Mathematics. McGraw-Hill, New York (2003)
141. AIM Mathematics test Years 3, 5, 7. Victorian Curriculum and Assessment Authority, Melbourne (2004)

Subject Index

Printed by Publishers' Graphics LLC
MO20120925-2159-174